Run from Easy

How to Run Towards a Self-Sufficient Lifestyle by Growing Your Own Veg Garden, Reducing Bills, Taming the Elements, & More!

Micah Bailey

A free gift to my readers!

Download My Free Self Sufficiency Planner At:
{ vaughanpublishing.activehosted.com/f/1 }

This planner is designed to help you track your self-sufficiency journey in a practical financial way and is completely free.

TABLE OF CONTENTS

Introduction   8

Is This the Right Book for You?

Allow Me to Introduce Myself

Chapter 1
Leave the Easy Path   14

Why Be More Self-Sufficient?

The Different Levels of Self-Sufficiency

How Independent Do You Want to Be?

Self-Sufficiency in the 7 main areas of life

Chapter 2
Self-Sufficiency & Food: Vegetables   20

The Right Vegetable Garden for You

The Traditional Vegetable Patch

Greenhouses, Aeroponics, & Aquaponics

Hydroponics

In Gardening, Size Doesn't Matter

How to Create Your Vegetable Garden

Pest Control

Hydroponics for beginners

Harvesting and Collecting

Chapter 3
Self-Sufficiency & Food: Meat, Eggs, & More   48

Poultry Farming for Beginners

Farming for Eggs & Farming for Meat

Slaughtering for Meat

Basics of Cattle Farming

Dairy Farming & Dairy Processing

Beef Farming

Ducks & Pigs

Sheep & Goats

Fish Farming

Bee-Keeping

How to Make Your Own Flour

Chapter 4
Food Storage & Preservation   66

Why Does Food Spoil?

Light, Oxygen & Temperature

Preserving Your Produce

Canning & Dehydrating

Salting or Curing

Pickling

Making the Most of Storage – Start a Pantry

Chapter 5

Taming the elements 78

Solar Power 101: How Solar
Power Works

Factors Affecting Efficiency

Solar Panel Aesthetics

Costs and Coverage

Wind: How Wind Turbines Work

Installing & Maintaining a Wind
Turbine

Water: How Hydro Electricity
Works

Installation and Maintenance

How to build a Water Wheel

Reducing Your Energy
Consumption

Efficient Home Heating

Reducing Heat Loss

Back-Ups and Generators

What Exactly is an Anaerobic
Digestion Biofuel Generator?

Converting Biogas to Biofuel

Beyond the Homefront –
Alternate Energy

Chapter 6

Water Self-Sufficiency 110

Harvesting Water from Streams

Harvesting Rainwater

How to Drill a Well

Reusing Water and Gray Water

How to Purify Water

Chapter 7

Waste Management & 128
Composting

Rethinking Basic Sanitation

What is a Composting Toilet?

Composting 101

Hot Composting

Worm Composting

Composting and Anaerobic
Bioreactors

Chapter 8

Do It Yourself 140

New Skill Resources

DIY Raw Materials

Skills Worth Learning -
Bushcraft

Some DIY Projects to Try

Raised Garden Beds

A PVC Pipe Vertical Garden

A Chicken Coop

Making a Composting Toilet

Cleaning Products

How to Make Soap

Shopping & Clothing

Conclusion 157

References 160

Introduction

If you rocketed into the twenty-first century as a time traveler from the past, you would be surprised by how much has changed. The onset of new technology has allowed for great social advancements and the ability to be connected to anyone on the planet instantly. We can just jump on a plane as if it were a bus, and travel 1000s of miles with ease. If we don't know something, we can just pull the information we require directly from the cloud and have it at our own convenience. There are many upsides to these advancements in culture as it has caused innumerous innovations to burst forth in all areas of life. However, socially, somehow many feel more disconnected than ever. Furthermore, business culture is struggling to keep up with the growing demand for the increased efficiency and productivity that the new technologies afford. This then breeds toxic work environments and stressed-out employees.

Many of us on the ground are just trying to get by in the rat race we call life with its various trials and challenges. Whether it's an ever-growing pile of bills that never disappears despite your best efforts, debt, or a stressful lifestyle that increases your heart rate at the cost of your free time. On top of this, we are more disconnected from nature, either because we live in the city or because we don't have time to go out and explore it; all the while we are growing in the realization that our toxic way of living is affecting the planet negatively. Simply put, life is hard. Now, more than ever, we are overstimulated and overworked, trying to plate spin much more than the average person had to 100 years ago. However, it doesn't have to be this hard. Have you ever imagined yourself with more time, space, and resources on hand? Many of us have, and yet, for some reason, we never achieve these goals.

I'm here to tell you that not only are you wrong about how much it costs to lead a self-sufficient and eco-friendly lifestyle, but that you don't have to be a conspiracy theorist to want to live a little off the grid. The key to freedom is not about getting more money and more possessions. If this is your means to free you from the system and culture we are living in, it will only make you more dependent on it. I believe self-sufficiency is a key to counteracting this unprecedented change in society. I will show you how to evolve your everyday into a self-sufficient lifestyle that will free up your time, money, and sanity. Pulling back from the culture of today will help you find more freedom than being rooted in the middle of it.

This book will show you what you need to begin your journey in self-sufficiency and see where your bills can be reduced and swapped out for self sufficient money generating/saving activities. Take hold again of your choice. Don't stay in a toxic job you hate just to pay the bills. Let the reason you work a 9-5 job be because you love it and enjoy doing it. Let it be your choice to be there, not a necessity.

Is This the Right Book for You?

If you are looking for a way to reduce your bills, lower your stress, and live a green existence, then in a word – yes. This book will teach you everything you need to know to begin revolutionizing your life, allowing you and your family to be self-sufficient in every sense of the word. With my help, you will learn how to save money on various utilities (including water and electricity), how to cut down your grocery store bills, as well as introduce you to one of the most underrated skills on the planet: DIY. Of course, you will also discover the wonders of growing and preserving your own food.

By taking back control of your lifestyle, you will be doing more than simply saving money and being a better friend to the environment. You will also be discovering skills and tools you never realized you had! Throughout this book, I will cover various aspects that will ensure you can successfully and comfortably live off the grid in certain areas of life. In fact, this book will focus on self-sufficiency in the different areas of life, with each chapter dedicated to another area of interest, including food production and storage, heating and electricity, DIY projects to reduce purchases, sewerage, and water treatment.

Allow Me to Introduce Myself

With everything I will be covering in this book, you are likely wondering what makes me qualified to share this information with you. This is the part where I tell you I am some professional or an expert in off grid living.

Off grid living is not something people really specialize in or take a degree in. One might specialize in the different areas that make up this book such as vegetable gardening, farming cattle, waste management etc., but usually trial & error, with a combination of experience are the best teachers and qualifications in this subject matter. My forte lies in the renewable energy systems discussed later in the book as I am a Mechanical Design Engineer by trade. I have also worked for multiple landscape gardening companies which has given me a good understanding of garden design.

Depending on your life experience, whether you are a homeowner, or whether you still have the travel bug, you may or may not desire to be completely off the grid. However, you would like to be able to live, at least in part, off the land. I am the same as you, and as a result, have a keen interest in the topic. I think we all wish we could be a little less dependent on a job for food and sustenance. We all would love to live off the land if we could but most of us are either limited by the space we own or live in, or are deterred by the difficulty of the task while holding down a job.

This book, I hope, will show you that while self-sufficiency has many different areas and topics which we could delve into, there are some key ones which will help you begin to feel the benefits of independent living while figuring out the level at which you wish to incorporate this into your life.

If you are anywhere near as excited as I am, then I won't waste any more of your time. Let's get started and change the way you think. I believe changing the way you think will inevitably lead to action; and it's time to take the first step.

CHAPTER 1: LEAVE THE EASY PATH

Find the less travelled road of self sufficiency

Before you start your exciting journey along the path less travelled, I would like to begin by clarifying exactly what I mean by 'self-sufficiency.' Basically, self-sufficiency is the antithesis of the modern concept of instant gratification or the culture of convenience. Today, it seems as though we have fewer skills than our forefathers had, and as a result, we are losing our ability to truly do things ourselves. Currently, most people would rather call in the AA than change their own tire, or call an electrician instead of learning how to change a fuse, or hire a random person to pick up their fast-food order instead of getting it themselves…

I am certainly not passing judgement on this kind of lifestyle, because we all do it, and outsourcing skills is a currency all on its own. The downside is that it has led to a generation of people who expect things to be easy to do, or even easier to get someone else to do. This might seem like a great way to focus on what is important to you while directly investing in the skills of others, but don't be fooled. Convenience culture can also be toxic, as it makes us totally dependent on others and on the system we create.

If you do not consider yourself a victim of convenience culture, ask yourself the following question. If the world as we knew it ended tomorrow and you found yourself alone, how would you survive? Would you break into shops and hunt for food that way or would find a plot of land and put it to work? Many of us cannot admit that we would have the skills to farm and fish and exist without external stimulus, and that's why this book is so important. Being self-sufficient does not make you a doomsday prepper or a social outcast. In helping you to become self-sufficient, my goal is to re-empower you to make the most of what you have, while learning how to improve the skills, tools, and resources you have at your disposal. In a modern world, self-sufficiency means survival, independence and more freedom.

Why Be More Self-Sufficient?

Self-sufficiency is just another term for independence, and there are endless reasons to be independent. Whether you want to improve your diet and lifestyle, prepare for possible political or social upheaval, or revel in the satisfaction of completing a project by yourself or as a family, there are more reasons to try to deviate from the easy path than there are reasons to stay on it. Deciding to become more self-sufficient helps both the planet, and yourself.

Of course, not everyone who embraces this change is driven by political ideals or a desire to stick it to the man. For many families, becoming self-sufficient is a way to reconnect with nature, while living via your own means, giving you more time for one another. This includes freedom from working for a salary or having to go to the supermarket, or purchase gas for your car. This may sound like something out of a dream, something impossible to attain. But there are many so many people with success stories in it and many of them only started small, meaning anyone can begin.

The Different Levels of Self-Sufficiency

One of the most important aspects to realize when seeking a life off the grid is that you can do it in phases or to different degrees. Not everyone has the desire or the resources to go totally off the grid and live like a hermit in the woods. For others, this is a truly attractive option. As we get to each new section of this book and with each new topic we cover, I will explain the ways in which you can apply the skills you learn in various contexts, so that they conform to what you expect and desire for your life. Whether you are only seeking to partially reduce your carbon footprint and establish a veggie garden, or you are seeking complete independence that means you will be handling your own electricity and sewerage challenges, you will find something to help you on that journey in this book. With my help, you will be able to decide if you want to rely fully on commercialism, or whether you want to make and repair everything yourself. There are different levels and different extremes to independent living, and you do not need to start at the most extreme end of the spectrum.

I often suggest that people exercise patience while starting along this road. Any changes you make – even if they are low maintenance and basic, will still have a positive effect on your life, so you should never

feel pressured to attempt projects beyond your scope or comfort. However, if you are here for a total lifestyle transformation and plan to make drastic changes, even better!

The first step is to establish exactly what you want to change. Are you searching for a way to eliminate expensive fixed costs from your life, such as mortgage or car payments, or are you trying to eliminate consumerism, over shopping and reliance on the social systems in your life, including your job? Your needs must dictate your exact goal when it comes to being independent.

How Independent Do You Want to Be?

Before we begin in earnest, it is vital that you establish your goals. After all, without a clear destination, it is tough to start a journey of any kind. Begin by analyzing your budget. List your set monthly or annual expenses against your income, and see where your money goes every month. For example, entertainment subscription services or electricity to power household appliances are generally money-hoarding culprits. Once you have an idea of what your money flow looks like every month, decide how you want this to change. What can you cut, and what costs could you be covering by learning new skills, particularly those that I will cover in this book? Once you know what you no longer want or need in your life, it is easier than ever to start working towards the life you do want. For example, if you waste money on groceries, the chapters dealing with growing and storing your own food will help. Similarly, if you want to start reducing your electricity and heating bills, this book will cover many ways to lower these bills and generate clean, green energy yourself. Sometimes there are alternative sources and solutions which help you move toward your goals. For example, reducing a food bill can be done by shopping at Aldi or Lidl instead of Marks & Spencer's etc.

At this stage, I urge you to focus on your goals and not how to get there. Chances are that you are lacking the skills you require to meet your goals, whether this is gardening or basic cattle farming, but they will grow with time. Regardless of whether you implement every strategy listed in this book or only the simplest, you won't regret the investment. We need a starting point in which you have an idea of what you want to change. The 'how' part of this journey begins now.

Self-Sufficiency in the 7 main areas of life

The way this book is structured is that the 7 following chapters will deal with 7 major areas of life showing the ways you can either: become more self-sufficient in them, reduce bills or dependance on external resources for them, or how to generate income/savings from activities associated with them. The following chapters will deal with these 7 topics: Growing Food, Animal Husbandry & Other Foods, Storage & Preserving Food, Taming your Bills; Heating & Electricity, Water Independence, Waste Management, and finally, DIY, Shopping & Clothing.

Chapter 2: Self Sufficiency & Food

FRUIT & VEGETABLES

One of the simplest lifestyle changes which will put you well on your way to this goal is to grow your own food. By maintaining a vegetable patch, you will eliminate or reduce the need to purchase vegetables and fruit from the grocery store. You will also save the time and expense of getting to the store. However, an equally important benefit of this change is that you will have better control of your diet. With fresh produce available immediately throughout the year, you will begin to eat more vegetables, improving your diet. Finally, you will be learning new skills about gardening, including tips on pest control and how to maximize your yield. The ability to grow your own vegetables is an incredibly rewarding skill, and a necessity, if you wish to run from easy. By the end of this chapter, you will know which basic crops to grow, and how to plan and create the perfect gardening system to suit your needs.

The Right Vegetable Garden for You

Depending on the level of self-sufficiency you wish to attain, your gardening needs will vary. If you are just starting out and want to perfect a small-scale project before extending to something larger and more independent, you may consider raised beds for a few simple veggies. However, if you plan to totally replace the produce required for a large family or small community, you will need something that could bear large quantities of good-quality crop throughout the year, and the system would need to be accurate and predictable. In this case, it would likely call for a hydronic greenhouse system. Whatever the end goal, I would recommend you start small and expand once you are confident and successful in your crop raising. This journey starts with baby steps.

These days, the word 'gardening' no longer refers only to planting something in dirt. As technology has advanced, and new gardening needs and trends have emerged (consider, for example, microgreens or patio gardens), the practice of gardening has evolved. Now, there are ways to make the most of what you have, despite limitations like space and light. If you have the patience to learn a few new skills, you could turn any space into a successful garden, which can produce crops of ornamental plants to suit your needs. I will begin by listing the most common gardening systems.

N.B. There is another benefit to creating your own vegetable garden - you eliminate your family's reliance on modern agriculture, which helps the environment, as modern agricultural practices contribute to greenhouse gas formation and the destruction of land.

The Traditional Vegetable Patch

If you have sufficient outdoor space in a temperate environment and access to high-quality soil (or the ability to create your own by composting), chances are you'll opt for a traditional vegetable garden or patch, in which rows of vegetables are planted in beds of soil. These beds can be raised (which allows the plants more root space in aerated soil and prevents some pests from accessing the stems and leaves) or low beds. If the beds are not raised, the top layers of soil should be aerated to ensure ideal root growth of seedlings.

Raised container beds are also an option for properties or homes that do not have a garden, but which do have sufficient space to house beds such as a large patio or paved area. Here, garden beds can be built and maintained in areas that would otherwise not be effective areas for gardens. By simply building a box planter out of sleepers or blocks and filling it with topsoil, you are half way there.

Greenhouses

If you have the resources listed above, but have inadequate environmental conditions, or numerous threats such as pests and frost, a greenhouse provides an excellent alternative to a traditional vegetable garden). Greenhouses are generally pricey to erect, but, in the long-term, are an excellent investment as they serve as protection and climate-control for gardens. Greenhouses are also ideal for containing outdoor irrigation systems. They allow you to make the most of a limited space, as many have sufficient space for shelving and hooks to mount hanging planters.

Aeroponics

Aeroponics refers to growing your crops in an unconventional environment – rather than soil, your plants grow in air or a misted environment. No soil is involved! Basically, vegetable seeds are placed in a piece of foam in a plastic pot. One end of the pot (the 'root' side) is subjected to a nutrient mist, while the other (the stem side) is exposed to light – either natural or artificial. This method of gardening is an excellent way to get the most produce out of a small space, as it eliminates the need for beds, and plants can be stacked and grown vertically in tubes rather than horizontally. According to Modern Farmer (Barth, 2018), these systems are extremely water-efficient. In fact, a closed-loop aeroponic system uses up to 95 percent less water than plants grown in soil. A closed-loop system also means there will be no nutrient runoff.

On the down side, these systems require a sizable amount of electricity to power the pumps required to keep the plants in a mist. The misting itself also requires a certain level of expertise – you need to add nutrients in exact amounts or risk damaging your plants. However, if you use solar panels, it is possible for the system to be self-contained or self-sufficient.

The crops that work best in an aeroponic system include lettuce, strawberries, tomatoes, and herbs such as mint and basil.

Aquaponics

Aquaponics is a great system for anyone also wishing to grow their own fish for food, as it combines growing crops with fish-farming. The most popular fish to pair with an aquaponic system is tilapia, as these fish are well-suited for crowded environments, and can also easily adapt to changing water conditions. However, other types of fish or marine life, from snails and shrimp to goldfish, will also work in these systems.

These systems can either be directly combined, wherein the fish live in the water from which the plants grow, or by adding wastewater from fish farming to the system. Essentially, fish waste serves as a source of nutrients for the plants in the water. As the plants take up these nutrients, the water is purified. If you have a combined system of fish with plants in a single closed-loop, the fish waste is essentially purified from the water while the fish eats any pests that enter the water, protecting the plants. It is a perfect mutualistic relationship. Surprisingly, one of the most important aspects of aquaponic success is the presence of bacteria – however, by this I do not mean harmful pathogens. Instead, the breakdown of fish waste which purifies the water is the result of microbial activity called nitrification, in which ammonia in the waste is converted to a form of nitrogen that is suitable for plant uptake.

An advantage of aquaponics is the ability to also grow and breed fish alongside your crops, while ensuring that the crops (and fish) are organic and pest free. However, this system requires space to house your fish tanks, which will be expensive to install. Furthermore, the correct pH balance of water is essential to the survival of the plants and fish, and thus, a level of expertise is required. In addition, some countries such as the United Kingdom require permits for rearing fish for the table. In the case of the UK, this legislation refers to an aquaponics setup of any size, and the main reason for this is to prevent disease and to prevent alien species from escaping and subsequently entering the UK's existing waterways (Aquavision, 2021).

The crops best suited to an aquaponic system include those which require low nutrient input, such as lettuce, spinach, kale, okra, herbs, and spring onions.

Hydroponics

Hydroponics is essentially an aquaponic setup without the fish; plants are grown directly in a nutrient-rich liquid solution. It is an increasingly popular way to grow crops. It requires sufficient knowledge of plant nutrition, and, if indoors, will require specialist lighting equipment. At this stage, I will not go into details about hydroponic systems as there is a section dedicated to this method later in the chapter. For now, it is worth knowing that crops best suited to hydroponic systems include cabbages, spinach, and lettuce, as well as tomatoes, cucumbers, beans, broccoli, cauliflower, herbs, and bell peppers.

In Gardening, Size Doesn't Matter

If you think that the miniscule size of your garden is what is preventing you from growing your own crops, you are wrong! Now more than ever before, it is possible to maximize space for gardening. As I explained above, some systems such as aeroponics, allow for vertical gardening. This, combined with traditional space-saving techniques such as shelving and hanging baskets mean that you can

turn a space of any size into a productive vegetable patch. If you live in an apartment or have basically zero space, consider renting an allotment to grow your own vegetables on. You would need to work out if the saving on food covers the rent of the allotment which I would recommend using the self-sufficiency planner at the end of this book for. However, even if the financial incentive does not warrant the investment in time, it may still prove a worthwhile experience to live off vegetables you have grown while actively choosing a greener lifestyle than when relying on the shops for your produce.

How to Create Your Vegetable Garden

Whatever your chosen gardening system, getting started on creating your ideal vegetable garden will follow the same series of five steps.

Step 1: Create a Plan

As with any project, it is essential to plan your vegetable garden before you begin with implementation and construction. The preliminary stages of planning include selecting the method of gardening that will best meet your needs. To establish your needs, ask yourself the following questions:

1. How many people do you need to feed, both currently and going forward?
2. Will you be growing crops throughout the year?
3. What is your pest control situation?
4. How independent do you need to be?
5. How do you want the garden to look?

These questions will help you determine the size, scope and requirements of your future garden. During this stage, carefully measure your available space and be aware of any aesthetic elements you want to include or avoid. For example, your garden does not exclusively need to be a site of agriculture. You may decide to include seating areas, or a corner for children to play. When deciding where you want to plant your future crops, keep the plants' requirements in mind. By this I mean be aware of how much sunlight the area receives during the day, how easy it is to irrigate, whether the plants will be sheltered, and what pests are near – be it human, animal, or traditional insect pests. Once you have a vision of the space, I strongly encourage you to sketch it, with measurements to see how it looks before you start.

Step 2: Where to Plant

This step is crucial to the overall success of your garden, as it will vary depending on the layout and resources, as well as your needs. Basically, you need to decide what to grow, and what containers to grow them in. These two aspects of gardening are interdependent – your crop will dictate the type of container, and a specific container is generally best suited for a specific plant. For example, a strawberry pot is an ideal place to grow berries but would present issues if used to grow carrots.

There are three main kinds of gardening beds, each with its own advantages and disadvantages.

In-ground beds are dug directly into the ground so that the crops grow level with the soil surface. Although these beds might have borders, they are not any higher than the rest of the soil surrounding them. Such beds are considered the easiest way to garden, as apart from water and nutrient additives through fertilizers, not much else is needed to start producing crops. You will not require any building materials, and the soil is easy to prepare. However, you will need to ensure the soil has not been contaminated and that plants are protected from pests, which have easy access to your crops. These gardens are ideal for any fruit or vegetable but are particularly handy if you want to plant fruit trees, which will need access to vast amounts of soil for their roots.

Raised beds are basically in-ground beds with a raised border, so that the level at which plants grow is higher than surrounding soil. The raised soil allows for more room for plant roots. The beds can either be built directly on the soil, in which case they have no bottom, or they can refer to boxes with a bottom that can be placed anywhere. The borders of raised beds are most commonly crafted from wood, but can also be made of bricks, mortar, clay, or repurposed containers. To increase the aesthetic appeal of the containers people often create them out of concrete block and clad with in Porcelain or Natural stone to match the paving they have installed. Whatever the material, ensure that your raised beds have adequate drainage, and that you do not use anything which may have toxic ingredients, such as pressure-treated wood. These beds are ideal for minimizing pest interference and are a great idea for gardens with no soil to garden in, or which have inferior-quality or contaminated soil. They also ensure adequate drainage of excess water, and can be a friendlier options for anyone with mobility issues, as they eliminate the need to constantly bend down to work in the garden. Unfortunately, they are pricier and more difficult to install, as they require materials for the borders and soil to be imported or moved. These beds are ideal for planting softer fruits (including various berries) and virtually any vegetable.

A third option is to plant crops in containers. This refers to any container that is not a raised bed and has a bottom (with effective drainage). They can be made from any non-toxic material which will stand up to regular water and sunlight, and exposure to the elements, traditionally, plant pots are made from ceramic or recycled plastic. The most important aspect of container gardening is that you need to ensure your plants have adequate space for their roots, and this is often far greater than we imagine. Plants like tomato or pumpkin require at least 15 liters of soil each for adequate growth! While container gardening will allow you to plant crops in soil-free areas, they are also an ideal way of beautifying your garden, as you can include ornamental plants or just attractive containers. Containers are also mobile – whether to protect the plant from inclement weather or pests or just for the sake of changing up how your garden looks. On the downside, containers for plants are generally pricey, particularly if you want something that is more attractive than a traditional plain pot. In addition, the limited amount of soil in a container means that your crops will dry out faster, as the roots do not have access to the water table. As such, these plants require more regular watering and fertilization. Due to the popularity of urban patio gardening, many new varieties of mini vegetables have emerged on the market. These are basically genetically modified plants that ensure the size of the plant, along with the fruit or vegetable it produces, is better suited for small environments. If you prefer to plant full-sized heirlooms (which I could recommend as these will also produce viable seeds for later crops), you may want to consider including stakes or growing frames to your pots, to guide your plant growth and provide structure.

If any part of this process is confusing, feel free to search for online garden-planning applications, as these can assist you in sketching out your garden and help you to visualise what it will look like before you begin!

Step 3: What to Plant

Now comes the fun – deciding which vegetables to plant. Of course, the crops you choose will depend on a number of factors, including what your family likes to eat, the space you have available, the season, and what grows easily in the country/region you are in. If you are totally new to growing your own food and do not know where to start, consider choosing two or three of the ten most popular crops. Once you get the hang of gardening, you can branch out into even more variety.

The best crops for beginners are those that are low-maintenance and grow quickly. Below are 15 common crops that are ideal for beginners.

1. **Beetroot**. The plants grow quickly, while taking up little room. Their shallow side roots make them ideal crops for containers, and they are easy to harvest. These vegetables have been associated with decreasing blood pressure and improving blood flow, as they contain vitamin B9, manganese, potassium, iron, and vitamin C (Bjarnadottir, 2019).

2. **Lettuce, kale and spinach**. I group these together as these plants share a common physical structure and ideal climate.

They tend to have shallow roots and do better in partially shaded areas, making them ideal for containers in the house, or shelving in greenhouses. A word of warning for anyone planting lettuce; these plants are fast seeders and spread remarkably easily. Do not be surprised to see lettuce cropping up in others beds over subsequent growing seasons. Lettuce is high in vitamin A, while kale and spinach are rich in vitamin B. This, combined with their low fibre, low calorie, & high water content makes them ideal for any diet (Martin, 2019).

3. **Cabbage**. These plants are relatively compact and have shallow root systems, making them easy to grow in smaller spaces or containers. They are relatively hardy, particularly if you are willing to sacrifice an extra leaf or two when eating them, and are quick producers. Unfortunately, cabbages are beloved by pests like slugs and snails, so consider planting these in raised beds rather than in-ground beds. Alternatively, you can surround the beds with copper tape, which is a natural snail and slug deterrent. Cabbage is a great addition to any garden as it helps lower blood pressure and cholesterol and is high in vitamins C and K (Kubala, 2017).

4. **Tomatoes**. Tomatoes grow on sturdy vines with medium-depth roots, and if you stake the plants to give them structure, they are easy to grow anywhere. If you have the space, tomatoes are also well-suited to hanging pots, where their vines will trail down below them. Admittedly, this is

also quite a visually pleasing way to grow them. Tomatoes are an excellent vegetable to have if you work outdoors, as they contain lycopene, which has been known to help soothe and prevent sunburn. They also contain high levels of vitamin A, K and C. In fact, a single tomato serves as almost half of your daily recommended intake of Vitamin C (Sass, 2018).

5. **Potatoes**. This dietary staple is surprisingly easy to grow, and can be started from a potato that has started to root – say, that lonely, shriveled one in the corner of your vegetable rack. Although these plants are best suited to raised beds and deep containers, an in-garden bed will work if you 'earth up'. This means piling layers of soil above the plant leaves as soon as they sprout, maximizing yield and saving the plant the time of having to grow deeper into the soil. Potatoes are also ideal winter growers, as the hotter summer weather makes them vulnerable to rot and pests. In addition to being high in fibre, potatoes are also high in antioxidants.

6. **Peas**. Granted, most people have a love-hate relationship with peas, and they are a nuisance to shell. However, they are easy, predictable growers, and, if you enjoy eating them, are a must-have. Peas grow on vines that can be encouraged to grow on stakes or trellises, making them soil space savers, as they naturally try to become vertical gardens. Alternatively, they can be planted in hanging baskets, to encourage the tendrils to hang and bunch. Peas are an

excellent food to strengthen the immune system, as they contain vitamins C and E, as well as zinc (Sengupta, 2018).

7. **Radishes**. These are surprisingly easy to grow, like beetroot, they require minimal soil and space. They also grow quickly whilst being easy to harvest. With their abundance of antioxidants, as well as minerals such as calcium and potassium, radishes are great for anyone needing to lower their blood pressure or improve blood circulation.

8. **Carrots**. For exactly the same reasons mentioned above (and the fact that these are such healthy vegetables), carrots are great plants for beginners. In addition, they can be grown throughout the year or for longer growing seasons than other vegetables, as they have a broad temperature growing range. In addition to containing vitamin A, carrots also contain calcium.

9. **Zucchini (baby marrows, courgette) and squash**. These plants, like the leafy greens, have similar structures as well as needs, hence the grouping. All four are remarkably low-maintenance sprint-growers, producing large plants with leathery leaves that creep. They are high water, fiber content and anti-oxidants. Conversely to their benefits, they do require a relatively large amount of space as they sprawl out. Although they can be staked to grow vertically, they still require significant root space. A small tip: If you are not staking your plant and leaving the fruit to mature on the

ground, be wary of pests and rot – harsh temperatures and slugs are ruthless to any vegetables or fruit that mature while resting on soil.

10. **Onions**. Like potatoes, these are a dietary staple. Onions, like beetroot, have the perfect structure and design for small spaces or containers. The leaves can also be trimmed regularly and used as green onions. Like carrots, onions are forgiving crops that do well in varied climates. Furthermore, onions are one of the healthiest vegetables to eat. They have been proven to reduce inflammation and decrease cholesterol, and they contain antioxidants (Kubala, 2018).

11. **Eggplant (aubergine).** These vegetables are not only easy to grow, with relatively shallow root systems, but they offer many nutritional benefits. One of these is that they are incredibly high in iron, making them ideal for anyone suffering from anemia or as an option to supplement a vegetarian or vegan diet.

12. **Chili peppers.** Peppers are fast growers that do not require much space, and they are also quite resistant to pests, making them ideal to plant between other vegetables as a pest-deterrent. In addition, they are incredibly high in vitamin C, and contain many anti-oxidants.

13. **Garlic.** Like chilis, garlic are excellent pest-deterrents and thus have the dual role of being edible, while protecting its

neighbors from insects. Garlic has long been considered a superfood, and the fact that it is so easy to grow (being both water-wise and having a shallow root system) means it is ideal for any small vegetable garden. It boosts the immune system, has anti-inflammatory properties, and is full of various minerals and vitamins.

14. **Corn.** While corn does require vertical space to grow, with stalks reaching up to two meters height when mature, it is an incredibly useful and versatile plant, which can be used to making cornflour, popcorn, or just eaten on the cob. It has also been known to assist digestion, as it has a high fiber content and is full of a variety of B vitamins.

15. **Cucumbers.** These vegetables grow on large vines, and so require space, but they are so versatile. With a high-water content and surprisingly low fiber content, they are excellent at preventing dehydration or serving as a low-fibre diet food. They also contain a surprising amount of vitamin C and can be used in picking projects.

Step 4: The Right Soil

Before you start to plant, have a look at one of your growing mediums. If your soil is very compact and retains water, it's likely high in clay. High clay soil, together with soil that is high in chalk (lime), will inhibit plant growth and likely result in sub-standard produce. To rectify this, consider importing better soil and creating raised beds or using containers. Alternatively, you could consider options like hydroponics or aeroponics.

If you are using soil as a growing medium, ensure that you have a mixture that includes soil-based compost and topsoil. I highly recommend doing a pH test of the soil – as the pH affects the nutrient availability and uptake in plants; as such, plants have different pH needs, although most crops (and certainly then ten listed above) prefer neutral soil (at pH7). Neutral soil is first prize as it will not require tampering. However, if your soil is acidic or alkaline, you will need to rectify this.

Sulfur (usually in the form of aluminum sulfate) will help make an alkaline soil more acidic, and thus closer to neutral. Alternatively, if you have an acidic soil, you will need to add an alkaline element to get it neutral. Here, lime is ideal. It can be ground into a powder or crushed into small pellets to be mixed with soil.

This section is only a starting point for soil science and agronomy. We will not be covering in any detail crop rotations, how to bioengineer your plants to maximise your yield or anything of that nature in this book. Just know that when it comes increasing the productivity of your soil and yield at harvest, the three best places to start are one; on your local Facebook group with others who are further on in their self-sufficiency journey to get free tips and advice, two; YouTube where there is an abundance of information on the subject or if all else fails, three; call in an agronomist to help.

Step 5: Construction and Planting

At this stage, you are ready to officially begin to grow your own food and become a gardener. The best time to start creating a garden, if you are beginning from scratch, is in the fall or winter. This way, once you have cleared the various weeds and grasses, and prepared your soil, you will be ready to start planting in the spring. If necessary, you can also lay down weed netting or membrane at this stage. While helpful in weed management and prevention, this is an expensive option. I prefer to skip this lining, and simply weed occasionally, as it gives me a chance to regularly check up on the plants and watch their growth. If you are going to be using raised beds, build or purchase beds that will last, and that are tall enough to not have to bend over to work in. Numerous designs and construction tutorials for such beds are available online. If you start planting much before the spring, you may want to consider building or purchasing cold frames to shelter your plants from the cold. These are basically transparent boxes surrounding plants (not unlike mini greenhouses) to let in all of the light and none of the cold.

If you require a larger construction such as a greenhouse or growing tunnels, I would recommend calling in professionals for assistance. However, it is possible to construct these yourself, particularly if you do so on a small scale. Polytunnels require a simple archway as a frame on each side, and can be wrapped with large sheets of plastic, making them easy to construct at home from recycled materials, if you are not able to purchase a top-of-the-line version. Greenhouses, on the other hand, tend to be pricier, as they are often sturdier and more permanent structures. However, miniature versions made of polyester and canvas are also available making it possible to have greenhouses even in the smallest of gardens. Not only do these structures shelter your plants from bad weather and most pests, but they allow for a temperature-controlled environment that effectively elongates your growing season.

Pest Control

Once your seeds are in the ground and your plants start growing, it may seem that every insect and animal in the vicinity is out to get your garden. Just as there are a variety of gardening methods available today, the ways of controlling and preventing pests are diverse. However, when it comes to easy pest control that best suits a self-sufficient lifestyle, I find that natural pest control is best. By this I mean eliminating the use of chemicals and toxins, and thereby reducing the risks of damaging the ecosystem of which your garden is a part.

Physical Control Methods

To keep larger pests such as birds, deer, dogs, or children from getting into your garden and damaging plants, you may need physical control measures. This includes fences or gates to surround your various beds & gardens. These controls also include netting over fruit trees to deter birds, or mulching plant beds to discourage hares & squirrels from masticating all your precious plant leaves.

Biological Control Methods

Unfortunately, not all pests are as easy to deter as deer. For most gardeners, insects and disease pose a much larger problem. A great way to prevent these pests is through biological control measures, which essentially introduce another organism (notable, the pest's predator) to the system to naturally remove pests. For example, some marigolds have root secretions that deter pests which traditionally feed on and destroy plant roots. Planting these near your crops will naturally prevent burrowing insects from devouring the plant roots. To deter flying pets such as flies and wasps, plant wild garlic. The flowers attract bees and butterflies, while the leaves deter other insects.

Chemical Control Methods

Ideally, these methods should be your last resort, if mechanical and biological controls fail to deter pests. Chemical controls refer to organic and inorganic chemical mixtures which are used to deter or kill pests. If you purchase such mixes to use, ensure that you prepare and store these according to the manufacturer's instructions. If you would first like to try organic chemical mixes, here are a few I would recommend:

1. **Used coffee grounds**. These are lethal to ants, and act as a deterrent to many crawling and creeping insects. Scatter the grounds around the edges of your beds to deter insects.

2. **Mint and garlic.** A mixture of blended garlic cloves with mint leaves, along with cayenne pepper powder and a drop per 100mL of dishwashing liquid is all you need for a natural bug spray that is safe to spray on your crops. Simmer the ingredients together overnight before straining and allowing to cool, and then it is ready for use.

3. **Neem oil**. If you have small insects (including aphids, mites, and small caterpillars), you can kill these pests by misting your plants with a mixture of water and neem oil (in a concentration of ten drops per 50 mL water). The oil prevents pest growth and molting, and is thus more effective against young insects.

Hydroponics for beginners

Hydroponic systems are ideal for indoor or greenhouse gardens, as they flourish when they are part of a closed-loop system. As such, they conserve water and allow for nutrient-dense crops. If these systems are housed indoors or in a greenhouse, they also create an ideal microclimate for your plants that is conducive for year-round growing and harvesting. As such, this method is theoretically the best option for anyone seeking to grow their own food and stay off the grid. You can purchase or construct a hydroponic kit – both options have their own pros and cons, but both are pricey. Unfortunately, a hydroponic set up is an investment, so when purchasing, remember the mantra of 'buy once, cry once,' and try to buy the best quality you can afford. Other disadvantages of this kind of setup include that it requires near-constant monitoring, particularly if you experience electricity issues such as regular power outages. Additionally, these systems are attractive to water-based insects and microbes, including mosquitoes, which lay their eggs in water. Finally, as with other closed-loop gardening setups, if one of your plants becomes diseased, this can spread incredibly easily and rapidly to your entire crop.

Regardless of your set up, the two most important aspects of a successful hydroponic system are oxygen and nutrients. As such, ensuring that the roots have access to oxygen through bubble systems and air vents, as well as perfecting the balance of nutrient additives will almost immediately ensure success. In fact, by perfecting the balance of nutrients, you can vastly increase yields, by ensuring the plant has access to the perfect nutrients in the ideal amounts. This also helps your plants to grow faster, decreasing growing time.

The logistics of starting a hydroponic garden are beyond the scope of this book, but tutorials and information are widely available on the internet should you require further assistance.

Harvesting and Collecting

Remember that nature could already give you the advantage of ready-to-harvest food in the form of nuts, mushrooms, and berries. If you own a sizeable acreage or have immediate access to local forests or open lands, keep a lookout for things you can collect and store. In addition, you can collect the seeds of plants that you can later germinate in your garden, saving you the effort and expense of ordering seeds. I will discuss preserving and storage techniques in greater detail in a later chapter, but for now, try to imagine how you can preserve or repurpose what you grow, whether through salting, pickling, or creating jams and jellies. If your land grows nuts and berries wild, take note of where they are located, harvest them as regularly as possible and store them for later.

Growing your own food is an important step if you plan to retreat from a life that focuses on convenience. Once you establish a successful gardening patch or garden, the possibilities are endless, as you will have built up enough knowledge and expertise to eventually grow any crops you may need.

You will find the way you think begins to change. Next time you are in a local pub you might see some flowers growing in the beer garden and start thinking about nicking some of it's seeds for planting in your garden later or you might be out on your normal dog walk but now you are spotting all kinds plants and trees you never did before.

Gardening opens up a huge range of possibilities and is especially exciting when you a receive tangible crop at the end of it. Having fresh produce all the time will make you wonder why you settled for anything less. Even just the fact that you are growing in your backyard will encourage you to eat more fruits & vegetables than before to ensure they do not go to waste, diminishing the amount of chocolate and unhealthy snacks you eat. I find that making healthy fruit cakes such as carrot &

courgette cake, banana and walnut loaf, rum raisin & mango cake are a much healthier, low sugar option to biscuits and chocolate too.

In the next chapter we will look at the other food groups such as bread, meat and dairy and learn how we can best tackle independence from the grocery stores & minimise our purchasing of them.

CHAPTER 3:
SELF SUFFICIENCY & FOO

MEAT, DAIRY, EGGS & MORE!

||||||||||||||

There is more to food than vegetables, and while growing your own produce is the first step in the journey toward self-sufficiency, we cannot forget about other food groups. Unless you are a strict vegan and refuse to eat or use products derived from animals, your trips to the grocery store will likely include purchases of baked goods, meat, eggs, and milk, among others. In order to truly live off the grid, you will have to ensure that you have access to these products without needing to go to the store. For that reason, a basic rundown of dairy and poultry farming is essential to independent survival.

Again, your precise farming know-how and needs will be dictated by your lifestyle, as well as by how independent you envision your future. For the purposes of this book, I will assume that you wish to gradually work your way towards total independence. The information I am about to give you about animal husbandry and livestock is purely introductory – I am by no means a professional. As such, I will cover the basics of livestock farming, and provide a theoretical background to the subject. While livestock farming will provide you with the meat, eggs, and dairy for your diet, there are pros and cons to working with animals, and the practice is regulated by various rules and policies.

To farm animals, you need three important resources: space, time, and money. Naturally, you will not be able to house animals without sufficient space for them to roam, exercise, graze, and reproduce. But many first-time homesteaders forget that livestock farming also requires the effort of purchasing and maintaining animals, which can become pricey if you need to purchase feed or take your livestock to a veterinarian. You will also need to ensure that you have the time and knowledge to ensure that your animals are happy and healthy. For this reason, it is vital to start small. After all, herds can always grow later through breeding.

Poultry Farming for Beginners

Of the various types of livestock farmed today, poultry can be regarded as one of the easiest. This is due to their small size and rapid reproduction. In addition, birds are often easier to source, transport, and rehome than larger animals like sheep or cattle.

Farming for Eggs

On average, a laying hen lays an egg a day (Department of Agriculture, 2017). As such, for about a dozen eggs a day, you'll need roughly the same amount of mature laying hens. If you plan to farm only eggs, you will not need to own a rooster, as they will fertilize the eggs, giving you more chickens but no sunny-side-up treats.

When purchasing your hens, there are two main options. You can buy day-old pullet chicks which you can raise into hens. One benefit of buying them young is that they will grow up being familiar with you and the environment you are keeping them in. On the other hand, you will have to be patient with feeding them whilst waiting for them to grow to a mature egg-bearing age of approximately 18 weeks. In addition, chicks of either gender are not as hardy as many would imagine, and particularly in cooler weather, face the risk of dying, especially if they are not vaccinated. A better option would be to purchase young hens called pullets, which are at the egg-laying age of between 19 and 20 weeks. If you do decide to get pullets, it is vital that they are fully vaccinated before you bring them home, to protect them against diseases such as bird flu, particularly as this can be passed on to other animals and humans.

Most egg farmers recommend that hens only be allowed to lay eggs for a year before they are slaughtered. This may sound cruel, but the reason is that after this time, hens begin to lay fewer eggs before they cease producing altogether.

To ensure that your hens are at their peak, you should allow them to be exposed to at least 16 hours of light on a daily basis. This should preferably be sunlight in a free-range setting, but if you have to keep your hens in a shed or enclosure, artificial lighting is available for purchase from several outlets. In addition, adequate ventilation is essential to mitigate air-born diseases and to keep your chickens happy.

With regard to feeding, the better your input, the better your output. Essentially, you want to provide your hens with the best quality feed that you can afford. Feed should be easily available to your chickens at any time, and they should also always have access to clean water. Typically, hens consume around 100 grams of feed a day, depending on their size, so keep this in mind when calculating how much you will need.

Farming for Meat

To get the most bang for your buck when it comes to poultry, I would advise a combination of laying hens and broilers. However, broiler farming has a few more restrictions than simply keeping chickens (or other poultry) for eggs. While the best way to ensure happy poultry is to allow your birds to roam free, in a world of limited space and multiple predators, this is not always possible. As with laying chickens, feed should be readily available, although broilers require a diet with a higher protein content than layers, and as such, peas and soybeans should be added to their feed.

Slaughtering for Meat

Although slaughtering regulations vary with geography, all poultry farming shares the common attribute of requiring veterinary oversight when administering vaccinations and any antibiotics, although these practices are currently being reviewed across the world. In fact, in January 2022, the European Union will introduce new legislation that will effectively ban all routine antibiotic use, including antibiotics which are administered as preventative treatments (McDougal, 2020). There are other regulations surrounding chicken farming, specifically with regard to the number of chickens you own, as well as what you plan to do with the meat. In the US, if you are slaughtering fewer than 1000 birds a year, there is no need for inspections or permits, the same applies when using the meat for your own food. However, if you wish to sell or trade the meat, you may require a license (Ploetz, 2013).

Basics of Cattle Farming

If one of your goals is becoming totally independent, but can't live without beef, you may want to consider your own herd of cattle. Alternatively, if your fiancé or spouse won't allow it, I'd recommend bartering or trading some of your produce for beef. Given the benefits of having at least one milk-producer on site, the decision against keeping a cow should be carefully considered.

Dairy Farming

There is more to dairy farming than keeping a few cows and milking them twice a day, although this is part of it. As ruminants, cattle have unique digestive systems consisting of a four-chambered stomach which dictates their grazing eating habits.

Unlike chickens, cattle are unable to produce milk without having mated – the only way that a cow can lactate is after she has had a calf. Traditionally, heifers (teenage cows, if you will) will need to be two years old before they can carry a calf, and how much milk they produce is almost entirely dependent on the amount and quality of feed they get. Usually, milk production is highest immediately after calving, and peaks within eight weeks of the calf's birth. Thereafter, it decreases.

Overall, cows generally lactate for 300 days after calving, after which milk production essentially ceases. As such, many dairy farmers inseminate their cows to calf once a year, to ensure they produce milk constantly, with the exception of a 50-to-60-day dry period before the next calf. For the scope of this book, you will not need machinery for dairy farming, as you should be able to meet your family's needs simply milking of one dairy cow daily. However, there are risks associated with dairy farming. Proper sanitation is essential to prevent diseases in your cow, and treating raw milk before consumption is equally as important. In fact, raw cow's milk is illegal in many countries across the world.

Pasteurization is the process of treating raw milk to kill bacteria and prolong shelf life. It requires milk to be heated to 71.7°C for a minimum of 15 seconds, before being cooled to below 3°C. This can be done using a candy thermometer and double boiler and does not require fancy or expensive equipment.

Dairy Processing

If you have milk and cream, it can be used to make butter, yogurt, or cheese. Of the three, butter is by far the easiest and quickest dairy product to create. All you need is heavy cream, a little salt, and some iced water. Beat the cream in a mixer or food processor as you would to make whipped cream, but do not stop. Keep whipping until the cream separates – this usually takes around four minutes, depending on the quantity of cream you started with. At this stage, you will have a group of pale-yellow lumps in a pool of liquid – buttermilk. Now, pour the mixture into your iced water to further separate the butter and buttermilk. Pour the mixture through a sieve to remove the buttermilk, which you can use in baking, and squeeze any remaining liquid from the butter, thereafter, add a dash of salt to taste, and enjoy fresh, home-made butter.

To make yogurt, heat eight cups of pasteurized milk over medium high heat, stopping just before it boils – you can use a candy thermometer to monitor this. Remove the mixture from the heat and allow it to cool to approximately 44°C. Add the cooled mixture to a bowl containing a ninth cup of warm milk, stirring until it is smooth. Then cover the mixture and allow the yoghurt to set for a minimum of four hours (overnight is best). Keep in mind that the longer it sets, the thicker it will become.

Finally, to make cheese, you will need warm milk (either fresh from the dairy, or gently heated on the stovetop). Next, you will need to acidify the mixture. This can be done with store-bought additives, but I would recommend using apple cider vinegar. This will result in a runnier cheese, whereas cultures will allow you to produce a hard cheese. Thereafter, you will need a coagulant – called a rennet – which should

be purchased (you can make your own using fig tree sap or other plant products, but that is beyond the scope of this book). After a few moments, the milk mixture will start to gel, and the top layer will solidify into a curd. By cutting this, you allow air into the mixture below, which will solidify it. As such, the more cuts in your curd, the harder your final cheese will be. Next, stir the mixture again so that the curds and watery layers below mix, over a low heat, the longer you stir and heat this mix, the harder the cheese. Once you are satisfied with the texture, remove the mixture and strain the curds from the whey (watery mixture) through a cheesecloth. Try to work as quickly as possible, so that the curd pieces are still warm, as this will help them retain their final shape, and make your cheese less crumbly. Finally, salt and age your cheese to increase flavor and pungency, and to harden the texture.

Beef Farming

One of the most surprising benefits of beef farming is with regard to feed. Cattle can actually consume crop residue and grain that would otherwise be composted, making them ideal for a large homestead. Traditionally, beef farmers purchase cattle that have already been weaned from their mothers (who are in turn likely used as dairy cattle), at about 10 months old. Thereafter, they are put to pasture in the spring, when grasses and resources are their most plentiful. However, in the context of off-the-grid living, you will likely keep your cattle in an enclosure with access to a feed of your choice. The enclosures should provide the cattle with enough room to move about comfortably. For many professionals, this means at least 1.8 acres of space per animal.

Unfortunately, many countries implement strict regulations when it comes to slaughtering cattle, and in many places, culling cattle on your property is illegal and can only be done at an abattoir. In addition, some countries have regulations that allow home slaughtering, while banning the removal of any part of the carcass from the slaughter site, meaning that you cannot process the meat elsewhere. For that reason, ensure that you first check any local legislation and procedures for cattle farming before you even think of acquiring a cow for slaughter.

Other Livestock

With the many limitations imposed by cattle farming – particularly with regard to space and slaughtering, many homesteaders instead choose smaller livestock, such as sheep, goats or pigs. However, rearing any of these animals may still be ruled by legislature, and I would advise that you research this prior to purchasing any animals. However, if you are not restricted by legalities, additional information on these animals can be found below.

Ducks

Like chickens, ducks can be farmed for both their meat and eggs, which are generally larger than chicken eggs. In fact, Pekin ducks are actually some of the easiest animals to farm. Although they require slightly more feed every day than chickens, they are excellent foragers and frequently take responsibility for their own diet. In addition, they are a great way to keep insects such as slugs and snails away from your plants, as ducks like to eat insects.

Where chickens have an advantage over ducks is with regard to predators. Unfortunately, despite being bigger than chickens, ducks are slower, and thus more prone to be preyed upon by dogs and foxes. If you do decide to keep ducks, you should consider fencing and other methods of preventing predators from accessing the birds.

Pigs

Despite being smaller than their bovine cousins, pigs are actually far more sensitive to their surroundings and particularly to temperature fluctuations. For this reason, it's important to ensure that you have an enclosure for your pigs that has a roof to protect them from the elements, particularly as these animals can suffer from sunstroke, such as a barn or hut. However, if you take the barn route, you will need to ensure there is adequate light and ventilation, as pigs are prone to dietary upsets like diarrhea, which can lead to dehydration and death if not quickly remedied. At her peak, a sow can produce up to 12 piglets during her gestation period, which lasts just 130 days. This will also depend on feed, as a protein content of approximately 18% is necessary for pregnant and lactating sows. And they need far more food than most people imagine – a sow with piglets will consume

58

between six and 10 kilograms of feed a day, while requiring between 25 and 50 liters of water daily.

Sheep and Goats

Sheep and goats are ideal for anyone wanting to farm livestock without excessive space – generally, sheep require only a fraction of the 1.8 acres that cattle need to ensure optimal health. In fact, you can comfortably house five ewes and eight lambs per acre of land! However, they can also easily be raised in barns and pens, requiring 15 square feet per ewe, making them ideal for a back yard or smallholding. Goats require slightly more space, at approximately 20 square feet per adult animal, and a little more if they are active. Feeding sheep is also far easier than with pigs or cattle, as they can subsist purely on grazing in pastures, without the need to supplement their diets. However, if you do not have good pastures with a variety of grasses, supplementation is a must. With regard to goats, they can be raised only on grasses, but this is not the best way. Ideally, goats should be fed on alfalfa hay, and if you cannot grow it, I strongly recommend you purchase enough to keep your goats well fed – which will mean between one and two kilograms of hay daily.

If you decide on extreme self-sufficiency, sheep and some goats also offer the option of wool, which can be sheared and spun into yarn once a year. Additionally, goats can be milked for dairy products, particularly feta cheese, although not everyone is a fan of goat milk, which has a strong flavor. However, if this is not to your taste, it can be sold to produce income for the household. This income can then be invested in other aspects of your journey to self-sufficiency.

While the ideal situation would be to live off the food you grow on your own land, some aspects of independent living (such as purchasing or repairing clothing) will still likely require an income to take care of. Feel free to prove me wrong.

A word of warning; if you have multiple kinds of livestock, and especially anything else in addition to pigs, ensure they are housed in separate enclosures. Pigs are omnivorous, and, if hungry, have been known to attack and eat lambs and kids.

Fish Farming

If you decide to try aquaponics, you are already on your way to becoming a fish farmer! Either way, your breeding stock will be housed in an enclosure that can be built in an existing body of water or built on land. As long as you have a smaller, separate enclosure for breeding pairs, to prevent other fish eating the new-borns or eggs, you are all set! The one important thing to remember with fish farming is that water should be oxygenated, and temperature controlled, and that dead fish are prone to bacterial infections, so they need to be removed from the water as soon as possible to prevent disease outbreaks. The best breed of fish to keep in a freshwater enclosure are tilapia, as they are prolific breeders who have varied diets and few predators. They are also cheaper to purchase than breeds such as trout.

Bee-Keeping

One of the most important but often overlooked animals to farm are bees. Not only are these insects vital for pollination and fruit formation, but beekeeping allows you to have honey and beeswax on demand. The nuances of this type of farming are beyond the scope of this book, but I feel it is still worth mentioning in brief. To begin, you can purchase a swarm from a reputable supplier and add this to an existing hive, or you can try to capture a swarm and move them into your hive. To reduce hazards and time, I would recommend the former option. The hives you use should ideally be inspected or approved by a professional, as there are specific requirements for a colony to start making honey. In addition, many cities have regulations surrounding keeping bees, and many areas do not allow beekeeping in residential areas, although these rules generally do not apply to agricultural land. If you are unsure, it is a great idea to first find out if there is any local legislation that prevents you from keeping bees. Beekeeping requires a range of tools and protective clothing, as well as a theoretical knowledge of bee habits and practices. As with other forms of livestock rearing, beekeeping could also serve as an additional income, as honey and wax could also be sold.

One More Thing Before We Move On

Of course, this chapter has not been an exhaustive account of the various types of farming and food production – to go into depth on these topics would require an entire library! But for the purposes of starting an independent lifestyle, these basic tips will have you well on your way. Before we move on to the next aspect of independent living,

I want to mention one more food group: the dietary staple of carbohydrates. Bread and bread-products are probably one of the most popular food groups (despite their relative lack of nutritional value). In choosing the degree of self-sufficiency you wish to attain (from entry-level to extreme), you also have various options for baked goods. If you decide to grow wheat and corn in your garden, you have the option of grinding these to create flour or corn flour, to make your own dough from scratch. Alternatively, you could purchase flour but still bake your own breads, or simply continue buying bread from the store. However, there is a definite satisfaction in baking your own bread and I would suggest you try this at least once, rather than continue to buy bread.

How to Make Your Own Flour

It is actually easier than many people think to make your own flour. By growing and processing your own wheat, you also ensure that your flour is free of the commercial bleaches and additives that are present in most store-bought flour. All you need is wheat berries (the actual fruit of the wheat plant) and a mill. The mill can be manual or electric. Each has its benefits – electric mills are much faster, and will give you a flour with a more consistent texture, however, they do require electricity, and often, the turning of burrs will release heat, which can denature some of the nutrients present in wheat. Manual mills also have the advantage of being multi-purpose; for example, they can be used for oilier ingredients such as nuts, to make nut butter!

If you are making flour to bake bread (rather than tortillas, pizza or other wheat-based items) then hard white wheat will be the best option to use, as its higher gluten content makes it ideal for the vigorous kneading that many bread recipes require. Usually, wheat berries will yield almost double the amount of ground flour, so how much you grind ultimately depends on how much flour you require. I recommend grinding a half cup at a time, to ensure you don't make more than you need. If you are using an electric mill, ensure that the machine is already running before adding the wheat, to prevent any of the fruit jamming the mechanisms. For added health benefits, use your flour within 24 hours – storing the flour longer than this will cause loss of nutrients through oxidation and exposure to light.

How to Make Corn Flour

If you have decided to grow corn, you can also process this to make corn flour. Because corn flour is gluten free, it is ideal to use in gluten-free baking. To make corn flour, all you need is corn, water and a little lime powder. The lime is made from sedimentary rock that is high in calcium, and can be corrosive, so when cooking with it, ensure you use non-corrosive cookware.

To start, heat three liters of water in a large pot on the stove. As it begins to heat, add two tablespoons of lime and stir slowly until it dissolves. Once all the lime has dissolved, slowly add one kilogram of corn ears (either white or yellow, depending on your preference). Bring the mixture to the boil before reducing the heat to medium, and simmering the corn for 15 minutes. After removing the pot from the heat, you can either immediately start processing the corn, or you can allow the mixture to soak for an additional hour, to allow the lime extra time to break down the corn, resulting in a smoother flour.

Next, drain the mixture and remove the corn hulls before discarding these (they can be added to your compost heap). The corn now needs to be dried. To start, pat the kernels down with clean kitchen towels, before grinding the mixture. This can be done in a specialized grinder, a food processor, or – if you want to go old-school and also save electricity – by pestle and mortar. Once ground, this flour will last for up to three days, and if you do not plan to use it immediately, should be frozen in an air-tight container to retain freshness.

Furthermore, you could consider bartering something you do have (be it vegetables, honey or even milk or eggs) for fresh loaves from other families that have decided to become self-sufficient, or who own homesteads. As well as bartering, you can use money generated from sold goods to purchase other goods that are harder to produce on your property like flour, clothing etc.

Chapter 4: Food Storage & Preservation

Now that you have successfully managed to grow and produce your own food, you may be wondering what to do with the excess, or how to make your food last as long as possible. This is a key part of off-the-grid living, as being self-sufficient means you will eventually eliminate those trips to the grocery store, even in winter. But how will you ensure you still have enough to see you and your family through the colder months, when that vegetable garden is either dormant or just slowing down? It is much simpler than you thought – allow me to introduce you to a few concepts of food preservation.

In this chapter, I will cover the basics of food preservation, and mention the most common methods of doing so. This list is by no means exhaustive, but will nevertheless give you a firm understanding of the basics.

Why Does Food Spoil?

Before we cover the many ways to use up excess food and prevent spoilage, it is vital to understand what causes your food to go bad. Understanding the biology and mechanics behind spoiled food will not only help you to better grasp the various methods of preserving produce, but will also ensure that you never have to wonder if something is past its use-by date again. While spoilage is a direct result of microbial degradation (rotting as a result of bacteria, molds or yeasts), these micro-organisms are helped or hindered by three important factors.

Light

Surprisingly, one of the chief culprits of food spoilage is light, as it catalyzes several chemical reactions in food. In plants, light leads to the degradation of chlorophyll (the pigment that gives plants their green color) as well as carotenoids (which gives fruits and vegetables their orange and red hues), which leads to discoloration of many fruits and vegetables. However, it also causes the breakdown of riboflavin in milk products, discoloration of fresh meat and the oxidation of vitamin C, making fruit less nutritious. Simply ensuring that you store food in containers which are not transparent is enough to curb light damage.

Oxygen

When oxygen is readily available, it not only helps microbes to grow, but it can catalyze certain chemical reactions in food, and most notably in fats, which can lead to bad odors. As with light, oxygen can also cause several plant pigments to denature, leading to faded and less nutritious fruits and vegetables. A simple way to prevent food being exposed to oxygen is to vacuum seal it - such sealers can easily be purchased from a variety of sources.

Temperature

Temperature plays an important part in food preservation, as it affects bacterial growth. Mesophiles are a group of bacteria that thrive at room temperature (technically, they grow best at

temperature ranges of between 20 and 45 degrees Celsius) and for that reason, affect humans, whose body temperature falls within this range.

As such, not refrigerating or freezing food may cause mesophile colonies to spread in foods, which when ingested, can cause stomach upsets and far worse. To avoid these bacteria, it is important to properly cook raw food, and to store preserved food in cold storage.

Preserving Your Produce

Any produce you are lucky enough to harvest from your garden and livestock which is not immediately consumed should either be preserved, or composted. There are several benefits to food processing, including elongating the shelf -life of food, ensuring you have supplies throughout the year, and for emergency or back-up rations in the event of something happening to prevent you harvesting fruits and vegetables in the coming season (you never know when you might fall victim to drought or fire). There are several ways to preserve food, and these range from thermal processing (freezing or heating), dehydration, fermentation and pickling, chemical treatments, and irradiation. Below are some of the most common methods of preserving your food.

Canning

This refers to a more in-depth process than simply placing food in cans. In fact, the key aspect of canning is heat processing, and this practice dates back to the 1800s. Essentially, food is placed in jars or bottles, sealed, and heated in boiling water. This kills any microbes in the food (and particularly the mesophiles which affect human health). If done correctly, the heat ensures that oxygen escapes the space between the food surface and the bottle lid, creating a partial vacuum which ensures no additional microbes or oxygen can enter. This essentially makes the food good to store for up to two years! Food that you wish to can should always be properly cleaned before starting. Because canned goods can be stored at room temperature (provided that they are protected from light), it is one of the easiest and most convenient ways to preserve food for anyone seeking a self-sufficient lifestyle. There are three main kinds of canning.

Water Bath Canning

This method is most effective for any food that is acidic, such as fruit. Essentially, bottles or jars are boiled within a pot that is large enough to ensure that the jars are completely covered throughout the heating process, and have at least an inch of water above the lids. However, do not allow the jars to touch the bottom of the pot directly, as the rapid change in heat, combined with any movement from bubbling water, could shatter the glass. Instead, insert a silicon mat or tea towel between the jars and the base of the pan. The precise heat and length of canning will depend on the food you have chosen.

Steam Canning

This process is similar to water bathing, and is equally well-suited for acidic food. Essentially, jars are placed in a loosely sealed pot in which steam is created, without increasing pressure. That means the jars may have their bases in water, but are not completely submerged. Usually, the jars are heated until the water runs dry, but considering the varying lengths as a result of different amounts of water, I prefer to avoid this type of canning, as results can be less predictable.

Pressure Canning

This process combines the benefits of water bathing with the added advantage of pressure, which ensures you face absolutely no risk of botulism. This process is ideal for non-acidic food. Essentially, pressure canning can be used for almost any food – from meats and vegetables, to soups and vegetables. While dairy products could be pressure canned, I would strongly recommend you dehydrate dairy products instead, as it is much easier and more effective. However, the disadvantage of pressure canning is that you need specialist equipment. If you don't mind working in small batches, and spending more time on the process, you can get by with a regular pressure cooker.

Dehydrating

Drying food in order to prolong its shelf life is a practice that was incredibly popular during wartime, when military rations were almost exclusively freeze dried or dehydrated. It is a highly effective manner of preserving a variety of food, such as fruit (and particularly apples, berries, peaches and apricots, mango, and banana), vegetables (carrots, onions, peas, beans and tomatoes), dairy products, meat (fish, beef, chicken) and soups. All that you need to be aware of is that your food should be sliced or divided into sections of equal thickness, to ensure consistent dehydration.

Generally, food is dehydrated in air, a vacuum, or by inert gases or steam. An additional benefit of dehydration is that only water is removed from the food, leaving the dried product with a higher nutrient concentration. Dehydration is most effective when the right balance is struck between low humidity, low heat, and effective air circulation. There are five main ways to dehydrate food.

Food Dehydrators

These are machines that can be purchased for the sole purpose of food dehydration. Generally, they comprise a heating element and fan which allows air to leave the machine via vents. This is the most effective, yet priciest, manner of dehydrating food.

Using an Oven

An oven works excellently to dry foods, but unless your oven has a fan, it takes significantly longer than a dehydrator, and will require you to prop open the door to allow moisture to escape. Additionally, you need to ensure that the oven is on it's lowest setting, and even then, you risk cooking food rather than dehydrating it. This makes it a more costly approach.

Drying in The Sun

This method is the most environmentally friendly and cost effective, but depends on your climate. It's nearly impossible to air dry foods in winter in most locations, or in areas with high humidity. If you have the right climate to dry food in the sun, be prepared to be patient – it can take several days, and requires constant vigilance against pests and insects.

Air Drying

This method is similar to sun drying, but takes place indoors, in a well-ventilated room. However, this method of dehydration is only truly effective for smaller items with relatively low water contents, such as herbs, chilies and mushrooms. These items can be strung up or tied into small bundles, before being left to dry. As an added precaution against insects and dust, I would suggest enclosing them in brown paper bags, but ensure that the air can still circulate.

Using a Microwave

This method is similar to oven drying, but much quicker. However, given that microwaves lack sufficient air circulation, it is not the most effective way to dehydrate larger items with high water contents, and I would recommend only using the microwave to dry herbs and leafy vegetables. The downside to this type of drying is that food is susceptible to burning, as even on the lowest settings, microwaves are hotter than the air from an element in an oven.

Salting or Curing

Salting is actually a form of dehydrating, and it works incredibly well for meat products – just consider jerky or corned beef, two of the most famous forms of preserved meat. Surprisingly, 'salt' in this regard does not refer exclusively to sodium chloride or table salt – there are various types of salt used in preserving meats, which contain nitrogen in the form of nitrates and nitrites, which prevent fat from oxidizing or spoiling. Occasionally, sugar is also added to a salting mixture, for flavor and to assist in preservation. Any meat you wish to salt or cure should be lean, such as beef brisket, pork shoulder, loin or ham. It should also be cool – that is, below room temperature. There are two main ways to cure meat.

Brine Solutions

Brine is a liquid salt solution. After soaking your meat in the solution for roughly five minutes, allow the excess liquid to drain from the meat before suspending it from a line in a well-ventilated space for two to three days. It is a good idea to do this in a drying cupboard or fenced off space that will prevent insects such as flies from getting to the meat.

A Salt Rub

This method involves rubbing a dry salt mixture directly to the surface of the meat you want to cure, and follows the process above, replacing brine with salt that is not rinsed off before hanging the meat to dry. This technique requires roughly three cups of salt for each kilogram of meat you want to cure, and requires additional drying time – generally four to five days.

Pickling

This term refers to any foods which have been preserved in an acidic solution. Traditionally, the solution is vinegar-based and works best with apple cider or distilled vinegars, which tend to have a 5% acidity level. This method of preservation works well for fruits and vegetables like cucumber, beetroot, squash, radishes, tomatoes, peppers, berries, carrots, green beans, mushrooms and onions. One you have an equal mix of vinegar and water, add a quarter of the volume of salt, as well as spices of your choice. Unlike canning, pickling does not require any heat, and most foods can be pickled raw (with the exception of eggs, which are usually cooked before adding the vinegar solution).

Fermentation

This type of pickling relies on the breakdown of natural sugars in foods, and is often aided by the addition of salt water, which catalyzes sugar conversion to lactic acid, which then pickles the food, eliminating the need for vinegar. Many self-sufficient food preservers prefer fermentation to pickling, as it preserves more nutrients and enzymes as a result of the lactic acid, and can be used to create alcohol. In fact, one of the greatest advantages of pickled food is that it contains a high level of probiotics, making it ideal for good gut health, and to boost the immune system.

Making the Most of Storage – Start a Pantry

There is nothing more disheartening than preserving food only for this to spoil while being stored. For this reason, I think it is worth including a few basic tips for ensuring your pantry is up to the task.

In addition to the obvious requirement that your pantry (and all the shelves and corners within it) should be clean, ensure that it is also free of direct sunlight, making it dark and cool, while still having effective ventilation. Regularly check on the contents of your pantry to ensure that no containers are broken, leaking, or home to molds or insects. Also ensure that you keep the oldest items at the front, so that these can be used first (this will help you prevent that collection of 20-year-old cans that seem to manifest in so many pantries).

Finally, make sure that your pantry is free from any chemicals (this includes cleaning products) or household trash (including recycling) to minimize contamination or attracting pests.

As with any aspect of self-sufficient living, how and what you decide to preserve will depend on your family's needs, tastes, and resources. There is no point preserving something that you will not eat, no matter how easy the process. Similarly, if you do not have the space or time to preserve food, there is no point in buying expensive equipment or tools for preserving.

Chapter 5:
TAMING THE ELEMENTS

If growing your own food is the first step toward a self-sufficient lifestyle, then being able to tame the elements could be considered the peak of off-the-grid living. For many, the idea of generating your own electricity – or at least being in control of how much you use – seems unattainable. Yet astronomical heating and electric bills are often the main catalyst for many of us to change our lifestyles. If you are fed up with those mounting electric bills that never decrease despite your best attempts to be creative and reduce your electricity usage, this chapter will show you how you can you reduce those bills, and possibly nullify them altogether. Alternatively, you consider generating your own power for the benefit of the environment, in order to live a green, eco-friendly lifestyle.

It is worth mentioning again that the degree to which you wish to be self-sufficient will dictate how independent you become with regard to electricity generation. By that, I mean that, as with all other aspects of off-the-grid living, there are various levels of taming the elements. Regardless of whether you want to reduce a few bills, become just partially independent, or embrace full-on off-the-grid-living with no dependence on the government or state power, you will find tips for how to start your unique journey below.

Before you can start the process of generating your own power, it is worth knowing which system or method of electricity generation will best suit your needs, resources, and budget.

Solar Power 101

Of the various alternative methods of generating electricity, solar power is not only the most popular, but the most adaptable and affordable, as the energy which it transforms is free! A solar-powered setup will work for a home or property of any size, as panels (semiconductor systems which convert sunlight into electricity) can be installed on the roof of any building, making this a neat and aesthetic option, too.

The popularity of solar panels also means that not only are supplies and technicians easily accessible, but the costs have slowly decreased over the past decade or so, making them much more affordable.

By using solar energy, you also ensure that your power generation is consistent throughout the year, and not affected by seasonal changes, which is a consideration for other methods of electricity generation, such as wind turbines. Furthermore, solar electricity is completely free of harmful emissions and by-products, making it one of the most environmentally friendly methods of energy generation, which will go a long way to helping you and your family to reduce your carbon footprint.

How Solar Power Works

Contrary to popular belief, solar power generation via panels comprises of two separate processes – electricity generation, and heating. It is possible to set up your system in such a way that you can make use of one or both aspects.

Photovoltaic panels are used to generate electricity, through solar cells (batteries) that are connected together within the panel itself. Linking two or more panels allows you to create what is called a solar array which forms one big electrical circuit.

How much power a solar array produces (or wattage) is dependant on factors such as the size of panels, the number of them in the array, the quantity of electrons captured by the semiconductors, as well as the cell's voltage. By changing these aspects, you can influence the amount of power you generate.

Solar power installation also requires the installation of a large battery, which is charged using surplus green energy generated by the solar panels. During the night, when the sun isn't out, the home can run on the power stored in the battery, maximizing the efficiency of the system and further extending the use of the green energy generated by the system. Sometimes it is possible to set it up in a way that the battery can be recharged at cheaper, off-peak times (this would be useful when the solar array isn't producing surplus energy). You would then be able to sell any excess back to the grid to generate income in times of high demand in the region. Having one or more batteries will allow you to manage your electricity usage and bills more efficiently and will counteract any weather-related fluctuations associated with solar power. Some energy providers allow you to charge using more green energy from the grid rather than energy that has come from non-renewable sources. Drawing green energy from the gird when it is cheap is ideal both during the night and during the winter seasons when the daylight hours are reduced and overcast weather is increased.

Once your panels are installed and functional, you have a choice of whether to connect your set up to the national grid, or setting your system up as a completely stand-alone system. This decision will also be influenced by how independent you wish to be. Essentially, connecting your setup to the grid enables both your battery to be charged up by the grid if it needs to, as well as allowing your energy surplus to generate you income. That income could then be put towards other areas of your life that are less independent so as to offset your overall reliance on a salary for sustenance.

In a stand-alone system, your panels would not be connected to the grid but to a solar battery which stores the energy created. This system is pricier, but ideal for those seeking full independence, or for areas in which it is challenging or impossible to connect to the national grid, such as rural locations or islands etc. Keep in mind that it is also possible to have a hybrid system, in which you are connected to the national grid while also being connected to batteries – the best of both worlds. Thermal solar collectors use sunlight to heat water stored in pipes and cylinders, which is then used to heat your home. The panels look akin to photovoltaic panels, but instead of having numerous cells packed into a frame, it comprises of a series of pipes and tubes.

If you had a hybrid system installed of both thermal collectors and photo voltaic cells, you could use one for heating the home and water for showers/dishwashing etc and use one for electricity.

Factors Affecting Efficiency

Naturally, solar panels function best in ideal circumstances, and these vary depending on a number of factors. To ensure that your solar panels deliver peak performance, it is vital to consider their location, size and angle.

Depending on where on the globe your home is, the angle of your solar panels may need to be adjusted – for example, the sunlight that strikes a roof in the northern hemisphere of the globe will not strike a roof in the southern hemisphere is the exact same way, because the earth is a globe, and the curvature of its surface affects the refraction (bending) of light rays. Ideally, solar panels should be mounted in a location that ensures they can capture as much sunlight as possible throughout the day, and preferably from sunrise to sunset. This means that the roof to which you want to attach the panels should essentially be in full sun, and angled correctly, and the ideal angle is south-facing. If you install your solar system through a company, they will usually assess this for you.

This does not mean that if your roof isn't at the perfect angle that solar panels won't work. You can adjust the angle and layout of your panels through the addition of brackets and mounted frames.

The position of your roof is not the only aspect you need to consider – there is also the issue of how much weight it can bear, and how many panels its surface can accommodate. The larger your setup, the more panels you will require, and this means added weight on top of your roof. As GreenMatch explains, a 4kW system normally has 16 250-watt panels, which together weigh roughly 280 kg, and requires at least 29 square meters of space (Vekony, 2021). Each panel requires approximately one and a half square meters of space.

While it is possible to buy solar panels in a variety of shapes and sizes, giving you some wiggle room when it comes to the surface area of your roof, larger panels do generate more electricity, and are thus the best option.

There is a common misconception that somehow solar panels produce less energy in cold weather. This is false. A solar panel will actually produce more energy on a cold bright day than on a hot bright day. However, due to the winter months generally having more overcast weather, the potential for a lower yield in winter still exist.

Finally, it is important to ensure that your panels are always clean and carefully maintained. Buildup of dust, leaves or bird waste will affect how much solar energy is absorbed by the cells in your panel. In the winter months, ensure that your panels remain clear of snow to ensure they continue performing.

Solar Panel Aesthetics

While some people like the modern aesthetic of solar panels on a home, others do not. For this reason, it is possible to purchase solar panels that also emphasize design. Some companies have even produced solar panels that look like French roofing tiles when viewed from the street. In addition, it is also possible to install free-standing solar panel systems, which can be laid in a garden, in a field, or even on a car.

Costs and coverage

Calculating how many panels you require will depend on your current (and projected) energy consumption versus your goal energy usage. For example, if you want to reduce your energy consumption, or if you require additional power for your new lifestyle, such as a hydroponic garden. Your roof size and the number of hours of available sunlight in your climate will also play a role in how much energy you can generate.

To calculate how much power your household consumes, you'll need those pesky utilities' bills, which will indicate your monthly consumption in kilowatt hours (kWh). Combine this with how much of that usage you want your solar setup to replace (from a fraction of the usage to 100% replacement) to calculate how many panels you will need, and how many your roof can accommodate. When you work with a company who installs solar panels, they will usually have a look at which side of your roof received how many hours of sunlight. My family have had solar panels installed recently and it comes with an app by which you can track your daily, monthly and yearly energy generation which also shows the monetary equivalent of what you have generated. This is ideal if you want to easily keep track of what you are generating, and it makes it easy to calculate the payback period of the investment: Total Cost of the installed solar array system/12*(average monthly income generated): Total number of years to payback. If you were to install a stand-alone system you wouldn't get these amenities and would have to calculate it manually like below:

Say the solar panel you want to install is a 250 W panel and you want to install 12 of them: Your system is therefore a 250W x 12: 3000W or a 3kW system. Per hour you could theoretically be generating 3kWh x 1 hour: 3kWh. Say you have 10 hours of sunlight per day and your panel is optimistically running at 50% efficiency on average throughout the day. 3kWh x 10 hours x 0.5 efficient: 15kWh generated per day. For 30 days, that is 450 kWh generated. To know what this looks like in money terms, the best way is to take your monthly electricity bill and divide the bill by the kWh used to see the price per kWh you were being charged. Then take that figure and multiply it by the monthly generated kWh. So, for example a medium sized house that runs solely on electric power (no gas), might have a bill of about £96/month which is about 1391 kWh per month. The price being paid per kWh is £96/1391= 6.9 pence. Therefore, if the solar power generates 360kWh per month, you are saving approximately 6.9p x 450kWh= £31.05 off your bill. The price for electricity depends on your provider.

Currently, a domestic solar setup can cost anything between £5,000 and £10,000 to install, depending on the size of your setup and your needs. For smaller homes, with a family of between one and three people, a 3kW panel system will be ideal, and will set you back approximately £5,000. For a larger family of three to five people, you should consider a 4kW system, which generally costs around £6,000 to install. For a family of more than four people, or to power extra items like hydroponic gardens or grow lights for poultry, I would recommend that you install a 6kW system. This is the priciest option, starting at around £8000, but will ensure that all your needs are met.

To calculate the payback period for the 3kW system in the example above; £31 per month in saving x 12 months: £372 per year. A £5000, 3kW system would take £5000/£288 = 13.4 years to payback. Which is quite a long time. If you have a battery as part of the system, it will store surplus energy generated (for when the house is using less than the system is generating) so you can use it later which will minimise the wastage.

However, a 6kW system would generate theoretically double the 3kW system which would put the yearly saving at £744. The payback period then becomes £8000/£744 =10.75 years. The larger the system, the faster the payback period providing your monthly electricity usage warrants it. In the U.K, to encourage the adoption of the technology the government have two tariffs that enable you generate further income from the panels which will bring down the payback period further. The first tariff is the Generation tariff which is where homeowners are paid for what they generate regardless of whether they use it or export it. The second tariff which is the Export or Feed-in tariff is where you get paid for exporting your excess to the grid which is currently 5.24p per kWh. These tariffs in combination with grants enable the initial investment to be less heavy on the wallet and can often bring the payback period down to approximately 8-12 years. However, the more it is adopted the harder to be successful in your application for these tariffs. For example the Feed-in tariff stopped accepting new applicants in 2019.

The systems usually have a lifetime of 20-25 years which will give you approximately 10 years of profit making. Please note, 50% efficiency for a small system is very optimistic. Bigger systems tend to be more efficient smaller ones.

Wind

If you live in a windy enough area, such as along the coast or in a mountainous area, you could consider installing a wind turbine to generate electricity. Like solar options, these utilize a free natural resource and have zero emissions, making them an environmentally friendly option. A small wind turbine can help you reduce your electricity bills by between 50 and 90%, depending on your consumption and amount of wind in your immediate surroundings. One of the added advantages of a wind electric system is that you can also use the turbine for pumping water, much like an old-school windmill. This could be connected to your water troughs for livestock, or used to pump water that can be fed into your vegetable garden.

How Wind Turbines Work

Have you ever wondered what causes wind, and why some areas seem to be defined by near-constant gale-force blowing while others are virtually breeze-free? Wind is actually a by-product of the earth's surface being heated unevenly, causing air molecules to move around and differences in air pressure. Wind-powered turbines harness the kinetic energy of these moving air particles through their spinning blades. The spinning action turns a rotary motor, which then charges a battery cell, storing electricity. The reason that these blades and motors are mounted on long 'stalks' or towers is because wind pressure increases as you move vertically from the earth's surface. As with solar panels, your wind system can be connected to the local grid (proving the same benefits and shortcomings mentioned in the section above), or it can be off-grid.

Installing and Maintaining a Wind Turbine

If you have average annual wind speeds of between 14 and 22 kilometers per hour (four to six meters per second) then you have the perfect conditions to install a wind turbine. To clarify, a 'gentle breeze' as reported by most meteorologists as moving at between 12 and 19 kilometers per hour. With winds at that speed, a small turbine (which is generally considered to be approximately 50 kW, and ideal for home use) wind will generate between 60,000 and 170,000 kilowatt hours every year, which is enough to power several houses for a year (Facility Management, 2021). However, anything slower than this, and I would suggest a different means of generating electricity, such as solar or water.

Next select a site for the turbine – preferably far enough from your home as it is tall, and without trees or power lines in the vicinity, and on a secure, level foundation. Alternatively, you could consider a roof-mounted turbine, if your roof is large and strong enough. These setups cost upwards of £2000, making them somewhat cheaper options than solar. The price of a tower turbine will vary depending on the size, but usually cost approximately £7000. A word of warning: you will need to check whether your local government has any legislation regarding wind turbines, as this varies from country to country. In the United Kingdom, building a wind turbine requires planning permission.

Furthermore, there are a number of regulations surrounding their construction, including that the property cannot have an existing wind turbine, that the blades of the turbine be higher than 5 meters above ground level, that it cannot be built in or near a conservation site (Renewable Energy Hub, 2010).

Assuming there are no regulations against the installation, you can either have a contractor install a turbine, or – if you are familiar with the skills of an electrician and have the necessary tools and equipment to erect a very small wind tower you could attempt to install one yourself. You would need planning permission in any event. To calculate the payback period of wind turbine, it is much more complicated than solar power as you have to understand air flow and basic mechanics to calculate the theoretical maximum energy generating potential of the system. However, once you have worked out the average kWh your system is generating and the average efficiency of the system (~30-50%) you are good to go. The same would apply for a generator used in a hydroelectric system such as water wheel below. Wind turbines have a minimum and maximum wind speed rating that they operate within. Efficiency and Power output depend on many factors such as the style of turbine used, the angle of the blades to the wind, the size of the blades, the wind speed, the power rating of the generator, the gearbox ratio inside the housing etc.

Water

If you are lucky enough to have access to a body of water on your property, such as a river, stream or dam, then hydroelectric power systems are worth considering. In fact, it is one of the cheapest ways to generate electricity, and is becoming increasingly popular around the globe, contributing to nearly 20% of the world's electricity generation (Just Energy, 2021). Surprisingly, it is also one of the oldest means of generating power, as rudimentary hydroelectric systems were used by the ancient Greeks and Egyptians for pumping water and powering mills.

While the electricity generation itself is free of waste and emissions, hydroelectricity does depend on a body of water and existing infrastructure. As such, if you do not already have access to water, it may be necessary to construct a dam, which is both costly and a disturbance to the natural environment. Furthermore, hydroelectric systems pose challenges in summer and during drought, when many streams or dams may run dry and thus fail to generate power.

How Hydroelectricity Works

Hydroelectricity is generated in part by gravity! Streams and rivers that flow do so on a gradient, or incline, adding to the momentum with which the water moves. All streams and rivers flow downhill. This momentum is exactly what hydroelectric turbines harness to create energy. Essentially, the water's kinetic energy is used to power an underwater turbine, which looks like a wind turbine, with rotating blades that power a generator which stores the energy in a battery. For this reason, the stronger the flow or current of water, the greater the amount of energy produced.

Because it is site specific, this is a power option that will not work for everyone, and will be particularly challenging in an urban environment, where space and zoning exclude many properties from access to moving bodies of water they own. In fact, planning permissions for hydroelectric set-ups can be a nightmare, so I highly recommend you thoroughly research the legalities surrounding this before you consider building a dam. Furthermore, not all bodies of water will work, unless they have both a flow (water that is constantly moving) and a head (a height difference which assists momentum). An alternative option is installing a water wheel system which does not require a dam.

Installation and Maintenance

The costs of installing a hydroelectric system can vary greatly, depending on what resources you already have. If you need to build a dam from scratch, it can become an extremely expensive enterprise. However, if you simply need to have a turbine installed in a stream that you legally own and are permitted to use, then you are good to go. While hydroelectricity is reliable with an excellent lifespan of half a decade, it is the priciest option for alternative electricity generation. Generally, a 25kW turbine is sufficient to power a house or homestead. This setup will cost in the range of £16,000 – which is roughly £700 per kW of electricity (National Hydropower Association, 2010).

How to Build a Water Wheel

Even without a dam you can use a water wheel on a moving stream. This age-old device converts the kinetic energy of falling water into electricity. It comprises a wheel divided into several sections or segments, each containing blades or scoops which are arranged along the wheel's outside rim, forming a driving car.

To construct your wheel, first decide how large it should be, as well as how many segments it should contain. Usually, wheels contain between six and ten segments, and these require precise measurement and cutting to ensure they form a frame into which the blades can be inserted.

94

Next, join the spokes at the center, so that they radiate outwards like a sun. It is a good idea to reinforce the center join of the spokes with a metal or wooden disk onto which the spokes can be bolted.

Attach the spokes and reinforced center to the frame, so that the entire device resembles a bicycle wheel. Repeat the entire process so that you have two flat wheels, and then connect these to each other using metal rods to form a spacer between each set of spokes at the rim. You can fix buckets, scoops or paddles in these spaces, to finish the construction of the wheel itself. I would recommend paddles, as these can be made from a short length of straight planks, and do not require specialist equipment to mount, and do not need to be angled or bent.

Thereafter, you can mount it to a stand, and then lower it so that the bottom paddle is submerged in water in the correct water flow direction of the stream. Alternatively, a raised version will have a pipe and water flowing on to the wheel from above causing the rotary motion.

The three main factors that influence the efficiency of this system are the torque of the motor generator being turned, the speed and width of the flow and shape of your wheel. You must optimize the resistance of the motor with the speed of the stream. This can be done with a gearbox between the main rotating shaft of the wheel and the motor generator. Wider paddles will allow more water to hit it and likely turn faster. But you also must consider the weight of the system; if it is made out of steel, it will last longer but will be heavier and therefore require more force to turn.

Another tip is that if your stream is slow moving, you can dig out an off shoot from the stream at a steeper gradient to allow the water to speed up prior to reaching the water wheel. It could then exit through a pipe onto the wheel and back into the stream further downstream. The picture above demonstrates this.

You need a way of storing the energy in a battery which will then need to be wired into your home to be useful. Alternatively, you could have multiple batteries charging simultaneously which could then be collected and switched out for empty ones. However, you decide to do, it is likely a great DIY project to get stuck into.

Another interesting type of hydro power is the Archimedes screw hydro turbine which, although it is beyond on the scope of this book, is worth looking into if your land and stream conditions are appropriate for installing one.

Reducing Your Energy Consumption

Having discussed the various methods of alternative electricity generation, it is important to make the most of whatever setup you choose. Off-the-grid living is generally synonymous with a more frugal lifestyle, and it's no surprise why. If you can reduce your power consumption, and the bills and carbon footprint associated with this, why wouldn't you? If you are determined to adopt an extreme level of self-sufficiency, it's important to be aware of how much power you use, and embrace the many ways of reducing this. Alternatively, if you do not want or need to generate your own power, but just want to lower your bills and make your lifestyle more frugal, this section can help!

15 Ways to Reduce Your Energy Bills

There are many ways to lower your electric bill without having to drastically change your lifestyle, but below are 15 easy steps to take control of your power bill.

1. **Go green.** Of course, the easiest way to save electricity is to create your own, and if you are able to introduce a solar, wind or hydroelectric system, you've basically already won the battle against the bills. This has already been discussed in detail above.

2. **Keep the heat in.** One of the quickest ways to drive up a utility bill is through heat that is generated and lost, particularly in the cooler months. By ensuring your home is properly insulated, and installing a draught excluder, you will be able to save on your heating bills by minimizing waste. In addition, it is a great idea to bleed hot-water radiators, as any bubbles in the water will negatively affect the heating of your home, using more energy for less warmth. Finally, lowering the overall temperature of your home by as little as one degree can go a long way towards saving electricity, and the difference is subtle enough that most families won't notice it.

3. **Replace halogens with LEDs**. One of the most common users of electricity is also unavoidable – lighting. However, by replacing any

halogen globes with LED varieties, you can drastically reduce your electricity consumption. Interestingly, installing dimmer switches can also lower the amount of power you use when lighting up a room, as can turning off the lights when you are no longer in need of them- the days of a fully lit empty home are long gone!

4. Rethink your laundry. If you use a tumble dryer to dry your laundry, consider investing in a clothes line instead, as this will drastically lower your power usage. Also, by lowering the temperature of the wash cycle (or switching to a cold-cycle wash entirely) saves on your heating bill. Some more extreme self-sufficiency enthusiasts have created their own manual washing machine using a barrel and rope to save on electricity.

5. Trade the bath for a shower. To save on your water and heating bills, try showers instead of baths, and consider lowering the temperature of your geyser's thermostat by one degree. Not only can you reduce your water consumption by up to 40 liters per wash, but you'll lower the amount of water that needs to be heated, too. A water-saving shower head (which essentially lowers the water pressure) can help you save even more water!

6. Switch off unused appliances. This seems like straight-forward advice, but many of us forget to switch off and unplug devices we aren't using, such as cell phone chargers or television sets. Even in standby mode, these appliances still use a trickle of electricity that adds up over time. Even better, if you can eliminate using appliances altogether (like using a whisk instead of a food processor or washing your dishes by hand rather than using a dishwasher or even defrosting food overnight in your sink rather than microwave), you can save even more!

7. Embrace fire. By cooking your meals on a fire (preferably built with sustainable wood), you'll cut down the energy you need to power a stove and oven. A fireplace can also heat your home naturally without the need for electricity, and candles can offer an excellent alternative to overhead lighting. By embracing the heat-giving and lighting abilities of fire, you'll easily save on electricity.

8. Defrost your fridges and freezers. When ice builds up in these appliances, it's more than just an intrusion into your freezer space – it means more energy is needed to keep everything cool. By regularly defrosting your fridge and freezer, and ensuring that the seals function, you'll lower your energy usage.

9. Do not overfill your kettle. This is one of the most common mistakes that many families make – filling a kettle every time it is used. Unless you're about to make several cups of tea or coffee at once, there's no need for this – only boil as much water as you'll need, and your utilities bill will thank you.

10. Plan ahead, and install surge protectors. If you are ever victim to power cuts, you will know that surges in electricity can damage your appliances, often leading to expensive repairs or replacements.

To mitigate this, install surge-protector plus to your appliances to make sure you do not need to replace or repair anything unnecessarily.

11. Turn off the taps. By turning off taps, and particularly the hot water tap, while doing activities such as shaving or brushing your teeth, you will save water as well as the electricity required to heat that water.

12. Change your summer diet. If you are absolutely determined to reduce your electricity bill, consider changing your eating habits in warmer weather. Instead of cooking food, opt for meals that do not require oven or stove-top preparation, such as salads or wraps.

13. Adjust your fridge and freezer temperatures. Similarly to lowering your thermostat by one degree, changing the temperature of your fridge and freezer can save electricity without actually affecting your food.

14. Install a geyser timer. You will be able to save a great deal of electricity by only having your geyser on for certain periods of the day – if you consider that most of us bathe only in the morning or evenings, there is no real sense to have a geyser on all day. All that is required to make this change work effectively is to plan your bathing times with more care.

15. Use a fan rather than an air conditioner. This saves money in two ways; firstly, when purchasing the fan instead of an air con, and secondly through usage. Fans require only a fraction of the energy needed to power an air conditioner, and they also come with the added benefit of being easier to maintain.

Efficient Home Heating

One of the main culprits of excessive energy consumption in the modern home is heating. Of course, this means that effective home heating will help you cut down on electricity bills, but figuring out how to do that is more challenging than many people think. For that reason, I have included my five favorite tips to ensure your home stays toasty, without the price tag. Not included in this list is the option to use solar power to heat your home, as these details were covered in the section above.

1. Check your boiler. Most homes are heated by a water boiler, so the temperature setting here affects your home temperature. For that reason, be sure to check that the temperature is not too high – as this will leech unnecessary electricity to heat. Usually, a temperature of between 60°C and 70°C is ideal to heat your home – anything higher than this is just a waste!

2. Adjust the thermostat. For years, I believed that turning up the thermostat would heat my home faster, and I was wrong. The speed of heating is in no way affected by the temperature, so turning up the heat is just ensuring that the water in your boiler heats to a higher temperature, which uses more power. Instead, consider installing thermostatic radiator valves (TRVs), which allow you to adjust the temperature of individual radiators, so that you can easily (and cheaply) adjust the temperature in different rooms, and save money doing so, but turning the heat off in empty rooms!

3. Turn off the heat when you don't need it. Just as a thermostat will not heat a room any faster if you turn it higher, keeping a room at a constant temperature is not the most efficient way to use gas or electricity, as it does not factor in any heat lost from your home. If your home is not properly insulated, heat loss will mean your thermostat needs to do double the work – replacing the lost heat, and keeping the room at a specific temperature. The colder it is, the more heat is lost, and the more energy you will consume to keep it at a static temperature. Instead, turn off the heat when you do not need it on, such as while you are at work.

4. Replace outdated boilers. While this is an expensive overhaul, many old boilers are far less energy efficient than newer models, no matter what tips you try to implement to save energy. For that reason, it is worth investigating newer equipment if you have had your current boiler for longer than a decade.

5. Rethink heating technology. A boiler and thermostat aren't the only ways to heat your home; they are simply the most conventional. For example, storage heaters charge during off-peak electricity times, making them cheaper to run if you pay a time-of-use rate. You could also consider an air-source heat pump, which works somewhat like a wind turbine, by using air to heat your home. While these are environmentally friendly, they do come at a steep price – literally. They cost anywhere between £5000 and £8000 to install (Marcus, 2020). Ground source heat pumps are similar – they circulate air and antifreeze through pipes buried in your garden to harness natural heat. However, these are expensive and difficult to install.

Reducing Heat Loss

As I mentioned above, regardless of how you choose to heat your home, you will still have the problem of heat loss. For that reason, it is essential to ensure you do all you can to minimize heat escaping from your home. Below are a few simple ways to make sure your home is fully insulated against the cold, and prevent heat loss.

Effective Insulation

When most people think of insulation, they think of a roof. However, there are many ways to insulate your home, including the walls, doors and even windows. In addition to insulating your roof or loft, you should consider cavity wall insulation. As the name suggests, this is insulation added to the cavities in your walls, between brick layers. Not only is this a relatively cheap option, but it can be done both at the time of building, or later, with minimal fuss. Also, because it comes in large sheets, installation is quick and easy, making it ideal to install yourself, and save labor costs.

Your windows can also be insulated through double or triple glazing. Essentially, this means double or triple layers of window pane, which not only prevent heat loss, but also insulate your home against noise, and can reduce condensation inside the glass, preventing mild water damage to your window frames or surrounding walls.

Further insulation can be done by adding draught excluders to your doors and windows, and even your letterbox, if you have a mail slit in your door! Finally, do not forget about your geyser! By surrounding your geyser with insulation, you can reduce your electricity usage, as the insulation will minimize heat loss, while also lowering the amount of time required for the water to reach a desired temperature.

Back-Ups and Generators

Even if you have discovered and implemented the perfect system for generating electricity, it is always a good idea to have a back-up. It is impossible to know if and when you will unexpectedly be left without power, and to guard against this, I recommend a generator, and specifically an anaerobic digestion biofuel generator. This is also a good idea for families that do not want or need to be fully off-the-grid, but who would still like a plan B for when the national grid is down. Furthermore, it is more environmentally friendly and cheaper to run than a diesel or petrol generator, as it does not require these fuels to create electricity, nor release smoke and carbon dioxide as a result.

What Exactly Is an Anaerobic Digestion Biofuel Generator?

The name of this generator sounds far more complicated than the work it does. Essentially, anaerobic digestion refers to the breakdown of organic matter by specific bacteria, in an oxygen-free environment. That is to say, bacteria digest and decompose wastewater, manure or even food waste within a sealed vessel (called a reactor) that does not allow the entry of oxygen. As the bacteria digest the organic matter in the reactor, they release by-products of digestant (solid waste) and biogas. As you might expect, one of the key components of this biogas is methane gas, along with carbon dioxide, water vapor and a few other gases. This biogas is used to generate electricity in much the same way as other natural gases, or can be used for cooking or lighting, instead of natural gases such as propane. In addition, the gas could be used as biofuel, after it has undergone processing and compression, in which inert gases are removed and the product is refined. The solid waste that accompanies the biogas is also incredibly useful – it can be applied to vegetable gardens as fertilizer, or added to compost heaps, making the entire system of great benefit to independent living. In extreme cases of independent living, the digestate can even be converted into bioplastics, although that is beyond the scope of this book.

What Can Be Used in an Anaerobic Digestion Biofuel Generator?

Amazingly, these generators do not depend on a specific source of biowaste – you can combine several different kinds of waste to the reactor, in what is known as co-digestion. These materials can include manure from livestock, food waste and scraps that you do not wish to compost, crop residues after harvesting corn, as well as a variety of oils, greases and fats of animal and plant derivation (including grease obtained from restaurants!) The only solid material I would advise against using is wood or hay, and the lignin and cellulose contents of these products are incredibly high, and cannot be broken down by the microbes, making them less efficient fuel material. These generators are also incredibly efficient, as up to 60 percent of solids added to the reactor will eventually be converted into biogas. In practise, this means that for every kilogram of dry material you add, you should expect between 6 and 18 cubic feet of biogas (Scheckle, 2018).

How to Use a Biofuel Generator

It is possible to purchase a biofuel generator or make your own, depending on your needs and resources. Unfortunately, I will not be covering the details of how to build such a generator here. The unit includes a feeding tube for adding your chosen wastes to the reactor, as well as an exhaust to allow the biogas to escape into a collection tank, and an effluent outlet to remove the digestate.

The process begins when you mix your chosen biowastes with water into what is called a 'feedstock'. The feedstock should ideally contain between 90 and 98 percent water, with any solid waste shredded into coin-sized pieces. Thereafter, it is important to add a starter culture, which is a mixture of methane-producing bacteria called methanogens.

If you are using manure in your feedstock, there is no need to add a starter, as dung contains sufficient methanogens to successfully digest the biowaste.

One of the most important aspects of the biofuel generation process is temperature – as you would expect for organisms that live in manure, methanogens or mesophiles, and thus function best at body temperature. Then it is simply a matter of waiting for gas to be produced. The time it takes for your feedstock to be converted to biogas is known as the retention time, and this will vary depending on the exact ingredient used in the feedstock. Once the rate of gas production begins to taper off, it is a sign that the feedstock has been completely digested, and a fresh feedstock should be added.

Due to the nature of working with compressed and flammable gases, you should never operate a biofuel generator indoors, and particularly not in an unventilated area.

Converting Biogas to Biofuel

For anyone seeking to lower fuel costs, a biofuel generator has an added benefit: the biogas it releases can be made into biofuel, replacing your need to stop in at a service station if you have a natural gas vehicle (NGV). NGVs are becoming increasingly popular, and are now being manufactured by several companies, including Honda and Ford.

It is also possible to convert a gasoline- or diesel-dependent vehicle into an NGV. This is done by installing fuel storage cylinders, fuel lines, regulators and a fuel-air mixer. Thereafter, a switch installed in

your dashboard can effectively help you alternate between traditional fuel and biogas. However, this is an incredibly pricey conversion, and will set you back anywhere between £4000 and £8500 per vehicle, depending on the engine size and type (Clarke, 2018). Then again, once your vehicle has been converted, you will save on your running costs, making it an investment to seriously consider.

Beyond the Homefront – Alternate Energy

One of the most impressive aspects of green energy is how adaptable it can be. For instance, it is now possible to purchase electric cars, which are not only environmentally friendly with zero emission, but are far cheaper to run, as they simply require an electric charge, which as illustrated above, is entirely possible to create yourself through alternate systems like solar and hydroelectric setups.

Alternatively, there are companies which are willing to buy back surplus electricity, or exchange this to cover the running costs of an electric vehicle. For example, Ovo Energy has commenced a trial in which users can charge their electric vehicles before selling any surplus (which is stored in a series of batteries) back to the company in exchange for credit towards their vehicle, or as a cash return (OVO Energy, 2019).

While this is by no means an off-the-grid solution (on the contrary), it is an excellent way for families to maximise their electricity generation benefits, while minimizing their bills.

108

This helps manage electricity use. Even if you are not completely independent from the grid, you would likely be able to meet all your electricity requirements from solar/wind setups, while using a car vehicle-to-grid system to reduce your entire electricity bill to zero. This way, you are self-sufficient with regards to electricity, but not independent (as you would still rely on the grid infrastructure. However, this is still very beneficial both to you and the environment.

Of course, these technologies are still in their infancy, and it could be several years before such projects are mainstream, but the exciting thing is that the potential is there. Similarly, selecting or converting to a new system such as electric cars and biofuels can seem like a daunting and risky exercise, but in the long run, it can be a rewarding experience.

Water is essential for all life, and without access to sufficient water, an independent lifestyle is impossible. For that reason, it is important to understand the various sources of water that are available, as well as how to best utilize these. In this chapter, I will detail the various alternate methods of accessing water, as well as how to purify and filter this to guarantee that it is safe to use.

Harvesting Water from Streams

The most important thing to consider when collecting water from streams is that stagnant or standing water should be avoided at all costs, as this provides a comfortable environment for microbes and bacteria to breed. While this water can of course be purified, it is easier to instead focus on flowing water in streams and rivers. Furthermore, try to avoid free-flowing water in busy areas or those close to ablution facilities, such as near campsites. Instead, travel upstream, where the water will not likely be contaminated by soap runoff or waste. To collect this water, simply hold your container (whether this is a water bottle, cup, or tank) in the stream until it fills. That being said, collecting fast-flowing water is not totally without risk – if the banks of the stream are uneven or slippery, you risk falling in and being swept away by a strong current, so ensure that you have firm footing.

Uses of Harvested Stream Water

Unless you have properly purified and treated any water you collect from a stream, it is unadvisable to drink it. However, this water is excellent for irrigation, or to give to livestock. Generally, if animals have access to stream water, they will drink it without any of the health repercussions humans could face.

Regulations Surrounding Collection of Water from Streams

Unfortunately, there are several regulations surrounding the use of stream water, particularly if this water source is located on public property. However, these regulations generally only apply to a watercourse, which is a naturally occurring body of water that was not man-made. As such, this precludes wells and dams that have been constructed, here, regulations are determined by the water owner.

If you are a landowner and your property includes a stream or river, you are entitled to riparian rights, which essentially means that you are legally entitled to receive the flow of water (provided that you have not altered the natural flow through reservoirs or channels). Riparian rights allow private fishing in the water, and water extraction for domestic purposes.

However, there are limits to how much water you may extract on a daily basis; in the United Kingdom, this limit is 20m^3 or 20 000 liters per day (Millar, 2018). These regulations may be different in other countries or regions, and as such, it is worth researching the riparian rights in your area prior to using water from streams.

There is another aspect of riparian rights that is worth mentioning. In addition to allowing you to use the water on your property, these regulations also hold you responsible for maintaining the integrity of the water. This means that you will be required to maintain any river banks, and prevent erosion or damage to these areas. Furthermore, it means you must allow the free passage of any fish, and cannot under any circumstances alter the path or flow of the water. In essence, riparian rights preclude you from any activity which infringes on the rights of other water owners in the same stream.

Harvesting Rainwater

Rainwater collection or harvesting pretty much what the name suggests – collecting rain water for later use. To be effective, it is important to ensure the water remains as clean and fresh as possible. Often, that means having a collection tank into which the water is collected. Regardless of your needs, harvesting rain water is one of the easiest and cheapest methods to ensure you have access to fresh water. There are many benefits to the practise, and chief among them is that there is no limit to how much water you can collect on your own property, and that it is completely free. Harvesting rainwater can also be beneficial to the sewer system, as it reduces stormwater runoff. Using rainwater to irrigate gardens is also beneficial to your plants, as this water is completely chemical -free and untreated, which means there is no chlorine, fluoride or other chemicals that build up in plant roots and stunt plant growth. Finally, rainwater is a great alternative for plumbing requirements, particularly in water-scarce areas, where using drinking water from taps to flush toilets or shower is irresponsible or impossible. That being said, I would advise against drinking harvested rainwater that has not been purified. While this water is considered clean, it is not always potable, or suitable for ingestion. This is particularly true if it is collected from a roof and gutters that are not clean, where it could easily mix with debris or microbes.

While there is an equation that can help you determine how much water you will be able to collect in any given rain-shower, the amount of water you collect will ultimately depend on how much rain falls to the earth. Nevertheless, one inch (or 2.5 centimetres) of rainfall over approximately 100 square meters will yield roughly 2000 litres of water!

How to Collect Rainwater

There are different methods of collecting rain water, and your choice of system will depend on the amount of water you wish to collect and then store. Generally, the easiest way to collect rain water is by attaching a hose to the gutters of your roof and channelling this into a storage tank. The easiest and most basic method is to collect water in barrels. Here, a barrel is used as a storage tank, and water is channelled from your gutters via a downspout and hose into the barrel, which is located at the base of the gutter. This method is ideal for people who do not require large amounts of water, or for anyone without the space to install larger tanks. The barrels themselves can also be recycled, further lowering your carbon footprint. The downside of this system is that there is a limit to how much water you collect (regulated by the size of the barrel), and anything exceeding this will overflow and be wasted. However, if you create an overflow of the rainwater harvester to irrigate the garden so that any overflow water is not wasted. A second system, known as dry collection, functions along the same principles above, but with a much larger storage vessel. This means that water can be stored between rainfalls, allowing the hoses that channel the water to fully dry out between each collection, while there is still water in the vessel.

By contrast, a wet system relies on underground collection pipes rather than a single hose from your gutters. This way, water can be channelled from several or all of the gutters in your roof into vertical pipes which spill over into a tank once the water pressure and volume increases. While this means you can greatly increase the amount of water harvested, and have a tank further away from your house (for practical or aetheric reasons), it comes at the price and the inconvenience of having to install the piping.

Installing a Rainwater Collection System

The great thing about collecting rainwater is that any type or material of roof and any shape or form of gutter will work – there is no need to invest in specialist equipment. The first step in the journey to harvest rainwater is to ensure your roof is prepped for it. By this, I mean that it is worth investing the time to clean the roof, and add gutter filters or screens, which will prevent twigs, leaves and other debris from collecting in the gutters and being channeled into your storage tank. You also have the option to add a filter to the gutter downspouts, but this is not necessary – you can always filter stored water later, before using it. However, I would recommend that you install a device called a first-flush diverter, which allows some of the initial rain water (the first flush) to run off before entering the tank, to prevent dirty or contaminated roof water from entering. This is essential if you do not plan to clean your roof before every rainstorm, which is highly impractical.

The storage vessel you choose – no matter the size – should be fitted with an insect-proof valve, which will prevent any insects from entering the tank and breeding in it. If you have a larger-capacity storage tank, and plan to use rain water for crop irrigation or for your home plumbing, it is worth adding a pump and water level indicator, although this is not necessary for small-scale collection.

How to Drill a Well

If your home is stated on property with a garden or with space, then drilling a well is an excellent way to ensure that your family has access to water without the need for streams, dams, or even relying on the grid. What's more, if you are willing to consider a well, it is incredibly easy to drill one by yourself!

Before you start to plan your well, you will need to find out if your local government has any regulations regarding the installation of a well – this practise is considered illegal without a special permit in some suburban areas. Next, you will need to select the correct site for your well, which should be located away from existing structures and pipes or utility lines, as well as terrains like swamps or marshes. The location of your well should also, wherever possible, be upstream of rainwater flow, to ensure that harmful chemicals are not wished into it during excessive rainfall.

Generally, sandy areas or those with significant gravel deposits are the best sites for well construction. However, these areas are also often harder to drill, because sand is eroded rock, and therefore the chances of boulders or large rocks being located beneath the sand are high.

If you are in doubt about the water level in a particular area, have a look at the surrounding vegetation. Usually, more plants indicate more underground water. If you are in doubt, you could consult local groundwater maps, which will be available from city planners or county offices.

Thereafter, it is just a case of digging deep enough to penetrate the various layers of earth and reach the water table. Traditionally, the best quality below ground water is located in sandy layers of soil, and an average depth of between four and six meters is usually sufficient to reach this layer. Generally, digging a deep-enough well will require an auger with an extendable attachment.

An auger works like a drill – turning it in the ground will displace earth with each rotation. You will need to remove and empty the auger each time that it fills, and eventually you will have a large pile of earth, so keep this in mind while drilling – placing the discarded dirt too near the well opening could later result in a cave in.

While it is theoretically possible to dig a well with hand tools and augers, if your terrain is very rocky, or features clay, I would recommend specialist earth-moving equipment to save you time and effort, particularly if you need to dig deeper than 6 meters.

After you have drilled the well, you will need to undertake a process known as bailing. Essentially, this means removing the now dirty water located at the base of the well. This is done with a bailer, which is a hollow tube or rod that can be lowered into the newly created space until you have extracted all the dirty water. It may require several attempts at lowering and then emptying the bailer.

From here, the pump can be installed, and the casing built. There are several other ways to create a well on your property, including sludging (in which you bore into wet earth) and tunnelling, but I find the drilling method the easiest to undertake.

Components of a Well

Despite the various types of well that are available, including drilled, dug or driven wells, they all share common components.

These include the well hole, or what most people consider to be the well itself, and casing, which could be seen as the walls of the well, which literally give it structure. The above-ground wall is known as steining, while the covering around the actual well opening in the surface of the ground is known as the well curb. In the image above, the steining is the bricked bottom of the well, while the casing is the reinforcing wall in the hole itself, which is not visible from outside the well. Apart from the casing, the curb is one of the most important parts of a well, as it prevents anything falling into the opening. In addition to the curb, many wells also include a well cap, which is a covering that prevents dirt and debris from falling into the water below. Most wells will also include a pump to bring water up from below, as well as the vessel in which the water is hoisted.

The Advantages of a Well

There are many benefits to having a well on your property, but these are the top three.

1. Saving money. Despite an initial investment in the equipment required to construct a well (including the relevant pumps and filtration systems), having access to your own water will eliminate the need to purchase water from the state, and this will save you a good deal of money over time. In addition, some countries offer benefits including tax breaks to the owners of a well, so while you are checking the regulations surrounding well construction, it is worth finding out if you will qualify for any added benefits or even government grants.

2. Being independent. If off-the-grid living is your dream, then a well is the answer to solving your water issues. Furthermore, it can simply provide you with an alternative water source to save money.

3. It is eco-friendly. To use water from a well with a filtration system will ultimately be more beneficial to the environment, as it avoids the pollution associated with water purification, and provides water free of chemical treatments such as bleach or chlorine. As an added advantage, in most circumstances, well water has been reported to actually taste better than water from the tap.

The Downside of a Well

1. Initial costs. The most obvious disadvantage to building a well is that it requires an initial investment for the relevant components and construction, particularly if you are using machinery to dig the well.

2. Possible contaminants. Without the chemical treatment that state water undergoes, well water may taste better, but it also poses a risk of contamination and bacterial growth. However, this is easy to work around if you are willing and able to install a treatment system, but once again, there are costs involved.

3. Electricity requirements. If your well water is pumped up to the vessel, you have the added disadvantage of requiring electricity in order to use your well. However, with the methods of electricity generation listed in the previous chapter, this does not need to pose a significant challenge. However, if you are not planning to become extremely self-sufficient and therefore do not intend to supplement your electricity generation, the need to power a well pump will be problematic. Alternatively, you can go the traditional route and use a manual bucket pulley system not to use electricity.

The Legal Ramifications of a Well

There are surprisingly few regulations against constructing a well. In the UK, you will require planning permission to drill a well in a suburban area, but on a property that is considered rural, such as a homestead, no planning permission is required. In addition, you are legally allowed to extract up to 20 000 liters of water a day from your well, without the need to purchase a license - this legislature is similar to the riparian rights available to owners of rivers and streams.

Reusing Water and Gray Water

Regardless of whether you have access to fresh water or not, there may come a point in your life (particularly if you wish to reduce your water bills or if there is a drought) that you may need to conserve water, or even reuse it. This is where gray water comes in. The term refers to used water that is nevertheless relatively 'clean' – having been used for bathing, or washing laundry. By re-using this water, you can reduce your domestic water usage by between 50 and 80% (Australian Government, 2010).

How to Collect Gray Water

Before you can recycle or reuse water, you will need to collect it. Thankfully, this is surprisingly easy, and requires just a few small adjustments to your daily routine. The easiest way of doing this is to decant your used bath water into a container, or add a basin to your shower to collect the water. Other methods include connecting your washing machine's outlet pipe to a tank to collect the used water, or decanting dirty dish water from your sink. Other sources of gray water include your pasta water, which is safe to use to irrigate your garden, as it is free of the soaps and chemicals that may be present in bath or dish water.

It is important that any collected gray water be reused within 24 hours, as water that stands longer than this may require additional treatment, as it is at greater risk of further contamination. Furthermore, if any member of your family is ill, it is advisable not to use gray water, as hand washing or bathing while sick could introduce microbes to the water.

Using Gray Water

Gray water has a number of uses, particularly if you use organic and solvent-free soap. It is particularly helpful to use for toilet flushing, eliminating the need to flush otherwise clean water. All that is required to do this is to fill the toilet cistern with grey water between each use, and not allow the cistern to refill with tap water.

Gray water can also be used to wash vehicles, and provided that it is chemical- and microbe-free, can be used to irrigate gardens. However, if you decide to use gray water for irrigation, there are a few key factors to consider. Firstly, the water should only be applied to plant roots, and not to the growing plant parts. This excludes leafy vegetables such as lettuce, kale or spinach, as well as root vegetables, such as carrots, beetroot or turnips, as such plants are susceptible to taking up chemicals in the water. When harvesting plants irrigated with gray water, careful washing before cooking is essential. Gray water should not be used in irrigation systems that mist or spray plants – instead, it is best suited for drippers, which prevent the water coming into contact with the plant itself. Finally, beware of excessive use of gray water when irrigating, as the soap content of this water eventually changes the pH of your soil to be more alkaline, which can be detrimental to plant growth. Unless it has been treated, it is not advisable to consume gray water, and this includes feeding livestock or pets.

How to Purify Water

If you have collected water from a source such as a stream or well, or if you are reusing gray water, you may need to purify or treat water to make it safe for consumption. Depending on the source of water, and its intended use, there are various methods of treating and purifying this precious resource.

Purifying Rainwater

Although rainwater is considered clean, it is not sterile. As rain passes through the atmosphere, it picks up dust particles, pollen grains, various spores and even larger debris particles such as bird waste. Furthermore, if you collect rainwater in a tank that is not regularly cleaned, you can accidentally introduce algae and microbes to the water, especially if the tank is exposed to sunshine, as the warmth will encourage bacterial growth. For this reason, all rainwater should be purified before being consumed.

The first step in the purification of any water is filtration. Depending on your needs, this can vary from a mesh over the water inlet and outlet pipes, to the installation of a commercial filter. Regardless of the type of filter you install, sedimentation (the introduction of small particles of sand and other matter) is unavoidable. However, the finer your filter, the less sedimentation you will notice. Once you have removed sediment, the water can be treated by ultraviolet light through the addition of a UV filter, which will eliminate any bacteria and microbes, making the water safe for consumption. If you are still hesitant about the quality of the water, you can also add a carbon filter, although this is not needed when purifying rain water. You can also boil water before using it, to be sure that it is completely free of any harmful microbes.

Purifying Ground Water

This is a more complicated process than purifying rainwater, as groundwater can also contain runoff from streets, commercial premises, or various agricultural activities. As a result, this water is often discolored because it has also passed through soil. The process of purification is almost identical to that used for rainwater, although depending on where you live, it may also require chlorination or the addition of bleach to make this water potable. Keep in mind that in large doses, chlorine is toxic to humans and animals, particularly to anyone with thyroid issues. As such, care should be taken if you plan to chlorinate your water. Traditionally, chlorine is added in tablet form to heated water. The ideal temperature to effectively dissolve chlorine tablets is room temperature, or approximately 21 degrees Celsius, or higher. In urban areas, I would advise against attempting to purify ground water, as the chances of this water containing harmful substances are much higher than in rural areas.

Wastewater Treatment

Unlike gray water, wastewater (or black water) includes water that is not fit for human consumption, such as sewerage, industrial and/or agricultural effluent, and even street water runoff that is collected in storm drains. However, it is possible to treat such water in the event that no fresh water is available. I will mention the process only briefly, as a detailed explanation of water treatment is beyond the scope of this book. Primary wastewater treatment removes the majority of solid material from the water & is done via precipitation and screening through basins. Thereafter, carbon dioxide and dissolved organic matter is removed during secondary treatment. Tertiary water treatment removes nitrogen, heavy metals & phosphorous compounds from the water, along with any remaining microbes, through adding chlorine and denitrification.

After being treated, water still needs to be purified. On a smaller scale, such as domestic use, this can be done through boiling or exposure to UV light (including sunlight) although generally, chlorination is the best method of purifying this water.

Chapter 7: Waste Management and Composting

An independent lifestyle has numerous benefits, but unfortunately, the downside of off-the-grid living means that you are also completely responsible for the management of your family's waste, as well as that from any livestock you may have on your property. This chapter will address the various challenges associated with managing waste on your property.

Rethinking Basic Sanitation

Generally, when someone mentions the word 'toilet', most people will picture a standard bowl and cistern set up with a flushing mechanism that requires anything between three and 11 liters to clear any waste. However, with the increasing popularity of independent living, combined with the knowledge that water is a finite resource not to be wasted, it is now possible to have a toilet system that does not require water. Not only will such a setup drastically lower your utilities bills, but it is a far friendlier option for the environment. Composting toilets take this one step further, by converting waste to a soil-enriching compost that will improve the quality of any home-grown food, while eliminating the need to purchase compost to enrich your soil.

What is a Composting Toilet?

A composting toilet is just one element of a larger system of composting, which will also comprise a more traditional composting bin. Essentially, the toilet is a collection point for the raw materials which will later be converted to compost. These are available in two forms – those that separate waste, and those in which the entire composting process takes place within a single chamber. Each has advantages and disadvantages, but the principles of use are practically identical, and a composting toilet that separates waste is usually easier to operate, and often costs less than the other variety.

Human waste fluid comes under the category of blackwater, and where gray water is safe for reusing elsewhere in the home, this is not the case for blackwater. The reason for this is simple: human waste contains pathogens that are eliminated from the body. When human waste is broken down by microbial action and exposure to oxygen, the presence of pathogens among the other waste products such as methane are what cause the unpleasant odor associated with poor sanitation practices.

There are several ways to reduce these odors, including adding a cover material such as sawdust or salt, which traps odor and excess moisture. However, a better alternative is simply to separate the raw ingredients into dry and wet materials, and many composting toilets feature a separation bowl and urine diverter, to do just that.

Separator toilets can funnel urine away from the bowl into a nearby pit of approximately 50 centimeter depth, which is filled with gravel. Once here, nitrifying bacteria in the soil will automatically begin the process of breaking down the liquid, making this non-toxic. An alternative option is to collect the urine, in order to water this down and use it as a liquid fertilizer. The nitrogen content in urine makes it ideal for plant nutrition. However, this is a messier option that actually depends on the health of your family, as when we are ill, pathogens are passed from our system via urine and faeces, and introducing any pathogens to liquid fertilizer is a bad idea.

The solid wastes captured in the toilet will thus dry out far quicker than they would if the blackwater was mixed, making this easier to work with and to extract when the container is filled. Depending on the size of your family and other factors, you will likely only need to empty the bin once a month (WooWoo Waterless Toilets, 2021). However, you do not have to purchase a composting toilet. Amazingly, if you have the time and skills, it is possible to make one yourself. The Internet is filled with various tutorials and blueprints for creating a composting toilet, meaning that you can avoid the costs of installation by a professional company if you are willing to try it yourself. In fact, the final chapter in this book includes a simple tutorial for creating your own composting toilet.

Composting 101

There's more to composting than toilets. In fact, successful composting requires three main ingredients: water, along with an equal ratio of browns and greens. In these circumstances, 'browns' refer to dry organic components such as dead leaves, twigs, and wood chips or small branches. Once these plant parts perish, their color will change from green to brown and they will release carbon dioxide as the chlorophyll within them breaks down, giving the ingredients their name. As you might expect, 'greens' therefore include fresh organic matter such as grass clippings, fruit and vegetable peels and waste from food preparation, and (contrary to what the name suggests) used coffee grounds and tea bags. Greens are generally items that are high in nitrogen.

What to Compost

In case you find the concepts of browns and greens complicated (as it is difficult to determine the nitrogen content of something simply by looking at it), here is a handy list of items that are ideal additions to any compost heap.

- Fruits and vegetables.
- Coffee grounds and tea bags, as well as coffee filters.
- Eggshells.
- Nut shells.
- Paper (including newspaper and cardboard, which should ideally be shredded into smaller strips).
- Grass clippings, leaves, yard trimmings and plants (however, I recommend avoiding flowering weeds, as their seeds are hardy and may survive the composting process, leading to a crop of weeds when you add this to a garden).
- Natural fabrics such as pure cotton or wool.
- Wood chips, twigs and wood shavings.
- Wood ash.

Avoid Adding These Items to Your Compost

Generally, animal products should not be composted, purely as a prevention against the introduction of harmful pathogens, and to avoid unpleasant odors while these items are broken down. However, other items that should not be added to compost include:

- Coal or anthracite ash (as this can contain traces of metals which are detrimental to plant growth).
- Any products from a black walnut tree, as these contain chemical substances that may harm plants, and particularly seedlings.
- Any plant matter from diseased or chemically treated plants.
- Oil, grease or fat (as these produce an unpleasant odor when broken down, but also attract flies and rodents).
- Pet waste, particularly from cats, as this may contain viruses and other harmful pathogens (however, if you plan to use the compost on non-edible plants, and are prepared to wait a little longer, you could consider using animal waste in hot composting, as the increased temperature will kill harmful microbes).

How to Maintain a Compost Heap

There are different ways of composting, including composting small batches indoors, but all successful composting shares a few basic principles; chief among them is patience and ensuring your ingredients are effectively layered.

To start an external compost heap, start by selecting an outdoor space with easy access to water, but which is dry and shaded. You can either add the ingredients to a pile in your chosen area, or dig a pit into which you lower a bin containing your ingredients. Add your various brown and green items as they become available, and moisten any dry ingredients as you add them. If you are adding whole vegetables and fruit to the heap, bury these under a layer of approximately 30 centimeters of other ingredients, to minimize the smell as they break down, and to trap the heat given off during decomposition, as this will increase the rate of composting. If you are using a bin or pit to house your compost heap, you should cover it with a sheet of plastic or tarpaulin, to prevent moisture escaping and the heap from drying out.

Then you need to wait for everything to break down. This process can take anywhere between a few months and two years, depending on the ingredients and the climate. Turning your compost heap (or the container in which it is stored) is a way to speed up this process, by allowing extra oxygen into the mixture and encouraging the growth of bacterial decomposers. However, it is not strictly necessary.

Your compost will be ready for use once it is dark brown or black, with a uniform texture, and no smell.

Indoor composting works on the same principle, but on a much smaller scale, using a bin or container that can be purchased from any retailer, or built. As this requires smaller input, the process is completed in a much shorter time span of between two and five weeks.

However, by adding commercial composting bran and enzymes, you can speed this up even more, and prevent any odor.

Hot Composting

This process seeks to speed up the degradation process by maximizing microbial action in the heap. As bacteria break down the ingredients in a compost heap, heat is released as a by-product. By providing the ideal environment for microbes, their activity is increased and intensified, leading to faster conversion to compost. To hot compost, all that is required is an amendment to the ratio of greens and browns. As I mentioned above, a traditional compost heap has equal ratios of these ingredients, whereas in hot composting, significantly more browns (carbon dioxide-releasing ingredients) are required than greens – in fact, a ratio of 25:1 is ideal (Vanderlinden, 2019). To help the mixture get started, you could add a shovelful or so of already formed compost as an activator, but this is not essential.

During compost formation, keep an eye on the temperature of the pile. Ideally, a temperature of between 30 and 40 degrees Celsius is perfect. If the temperature dips below this range, it is time to turn the heap. Regularly turning the compost will ensure that this heat remains constant, and that it is evenly distributed.

Hot composting will yield an end product in as little as three weeks, so it is well worth the added effort of monitoring the heat and turning the pile regularly.

Worm Composting

As the name suggests, this process relies on worms, and specifically earthworms. Also known as worm farms, this system relies on earthworms to process vegetable scraps and other traditional compost ingredients into vermicompost, or worm compost. This is essentially worm waste, and it is not only high in nutrients, but in forms of nutrients that are easily absorbed by plant roots, making it a highly effective compost.

Despite the benefits of vermicompost, this type of composting poses even greater limits on raw ingredients. Essentially, a worm farm should only include fruit and vegetable scraps, as items such as ash, paper, coffee ground or cotton will not be digested by the worms. While animal products such as eggshells will work in a worm farm, these items are broken down very slowly and ineffectively by worms, so it is better to avoid them.

All your worm farm requires for composting success is a happy environment for these bugs – a container that shields them from the sun and provides a dark, warm and moist environment. However, air circulation is essential, so this cannot be a closed container. Your worms can rest on a bedding made from moist newspaper or grass, and you can add your fruit and vegetable scraps whenever you have them. Within three to five months (depending on the size of your bin), the compost will be ready, and the cycle can be restarted for the next batch.

Composting and Anaerobic Bioreactors

In chapter five, I mentioned an anaerobic bioreactor generator as a means of creating your own biogas. However, because these generators rely on biowaste, they also serve as a composter of sorts. In fact, because these generators can create biogas from almost all biowaste – including animal products and oils which are not suitable for composting, they are an ideal partner to a traditional compost heap, as a way of totally reusing your household waste. You could, for example, add whatever ingredients you cannot compost to the generator, create a closed system which is powered by waste. On the one hand, you will create biogas, and on the other, a rich compost to improve your garden. This is just one of the many ways that the various methods of self-sufficient living can be combined to make more extreme levels of self-sufficiency both more easily attainable and more efficient!

CHAPTER 8: DO IT YOURSELF

The self-sufficient lifestyle is a DIY lifestyle. With regard to independent living, 'do it yourself' goes beyond construction projects and repair, to learning and implementing new skills in a variety of disciplines. Whether this is making or fixing your own clothing, metal working, horticulture, or woodwork.

It's the very definition of self-sufficiency to be able to create something that you need or want, using skills and resources already available to you. This, like other aspects and practises discussed in this book, can also be adjusted according to the level of independence you prefer. Partially self-sufficient people might fix something themselves for the sake of saving on an expense – why buy when you can fix for free? This is especially important if the item or equipment you need to fix is unique to your lifestyle – such as a hydroponic gardening system.

Alternatively, you could generate additional income for whatever purpose by making and selling something. On the other hand, someone wishing to be completely self-sufficient could create or repair anything that might otherwise require them to buy goods and services, by making everything they need at home, from clothing to cleaning products or building renovations and furniture. Of course, these principals also all apply to the concept of trading and bartering.

Learning a skill means you save the time and cost of hiring someone else to perform that skill for you. This chapter will discuss several of the most important.

New Skill Resources

In this day and age, the Internet is a DIYer's greatest resource. There are sites that offer everything you need to know about how to learn a new skill. Whether you enroll in online classes that offer study and certification of a skill, to blog posts with blueprints, tips and videos – anything you want to learn can be found, and easily. If you are totally new to the concept of DIY, you can start with small simple projects to build your knowledge and confidence. Initially, you may still need to hire specialists to perform tasks that are dangerous or beyond your capabilities (such as rewiring your home), but eventually, you can work your way towards being able to do it alone.

A Word About Tools

The basic skills and projects we will cover in this chapter all require the same basic set of tools. They are the very ones you will most often need for general home maintenance, so it is a good idea to ensure you have everything. They are:

⇒ A claw-foot hammer. Every toolbox needs a hammer, but the claw is also great for pulling out nails and studs, or functioning as a level or mini crowbar.

⇒ Pliers. These are useful for cutting and stripping wires, bending small rods or wire, and gripping things or other tools in awkward spaces.

⇒ A drill. While it does not really matter whether you go electric or cordless, the latter are a better option. A cordless drill is far more compact and mobile, and with a few extra drill head attachments, can function as a multi-tool – a sander or polisher, grinder, or screwdriver.

⇒ A set each of flat- and Phillips-head screwdrivers. These will enable you to work with screws and nuts of most varieties. The Philips has the cross-shaped head. Also ensure that you get a few sets of basic nails and screws -there will never be a time when you do not need a screw or nail for some quick project or fix, even if it is just to hang a picture.

⇒ A set of wrenches. These are essential if you ever need to work with nuts and bolts or any kinds of pipe and pipe fittings.

⇒ Spirit levels. Some spirit levels come in a casing that can be used as a ruler, and that is doubly useful. These will help ensure that your lines – whether cut or drawn or painted, are straight, even if they are angled. For example, if you are installing shelves to a greenhouse wall, these would ensure the shelves are level, so nothing falls from them.

⇒ A tape measure. These are so helpful when planning projects, as nearly all DIY projects – from building a chicken coop to sowing a garden or mapping out a well – will require precision and careful measurement.

⇒ A ladder. Ideal if you need to get for your roof, to install rainwater harvesters or solar panels. A ladder is also helpful in the garden, for trimming trees or reaching wind turbines.

⇒ Various kinds of tape. These are handy for quick fixes and

143

finishes. I recommend duct tape, masking tape (painter's tape), as well as aluminum tape, which is excellent for any repairs to solar-powered equipment.

⇒ A shovel. This is useful for more than digging. It can also be used to mix cement, turn compost, and maintain vegetable garden bed edges.

⇒ A saw. If you plan to do any wood work or construction projects, this is essential. Furthermore, if you have access to trees (such as a woodland or forest) it might also be worth considering a chainsaw, which would make tree-felling much easier.

⇒ A utility knife. This is helpful for cutting boards, carpet, tape, plastic pipes or even aluminum sheeting.

⇒ Safety gear. This includes goggles, safety masks, gloves, earmuffs (especially if you have power tools or chainsaws). Also remember a fire extinguisher and of course, a first-aid kit.

DIY Raw Materials

If you plan to create, maintain and repair enough for your family to live in relative independence, you will require resources for the various skills discussed throughout this book. That is, for living off the land and taming the elements.

Wood

Unless you have access to a forest or have woodland on your property, you will likely need to purchase timber for any building or woodwork projects. However, there are also other ways to get wood for burning, building, or other projects.

Salvaged wood is not only free, but by recycling or upcycling this resource, you also help the environment. You can find salvaged wood near your home by doing an Internet search, or a look through the local community newspaper. If you have construction sites nearby, they might be willing to give or sell you wood in their dumpsters. This is ideal for larger pieces such as doors, window frames, beams and pillars. Similarly, you could approach farmers to ask whether they have any old timber from animal pens or sheds – often, they will be willing to give or sell these pieces. You could also approach shipping or packing companies for wooden pallets, because these can easily be repurposed into a number of projects, as we will see later in this chapter. There are also many antiques dealers and construction companies that sell salvaged wood at reasonable prices, and in this way, you could also get your hands on specific pieces, such as old railroad sleepers.

Building Supplies

Depending on the size and location of your home, you may also have access to natural resources which could be used in various projects, including construction. This includes clay, which can be used in building, or even creating ceramics; bamboo, which can be used for piping in gardens or even for poles or scaffolding for other projects. Boulders and rocks could be used in masonry projects. Most building supplies will likely require purchasing though.

Gardening and Agricultural Supplies

If you plan to start your own garden or keep livestock, you will need more than just the space to do so. Good quality soil encourages plant growth, which is necessary for feeding your family and any livestock. Be sure that before you start a vegetable garden or plant crops, that the soil has been assessed and, where necessary, been supplemented with compost or extra nutrition, or an additive to change the pH.

Skills Worth Learning - Bushcraft

Many of the basic skills you may need in an independent lifestyle fall under the heading of bushcraft. This refers to the process of using natural resources to survive, by using knowledge of how best to use bushcraft tools to tame these resources. Essentially, it requires the knowledge of various kinds of handcrafts. While this craft was previously in the domain of wilderness survival, many of the skills and practices are useful for modern independent living. In a nutshell, these skills include foraging for food, as well as hunting and trapping game, water gathering and purification, shelter building, fishing, and the best ways to light a fire. As such, this craft requires knowledge of various things from how to tie knots, to how to strip bark from a tree or create a water filter. Of course, if you do not have much land to work with, you will have no need for skills like hunting, but it may still be worth learning to prepare and clean fresh meat.

Some DIY Projects to Try

In this section, I will provide some basic tutorials for several of the projects recommended throughout the course of this book.

Raised Garden Beds

The easiest way to build a raised bed is to use planter blocks, which are cubes into which wooden planks are slotted. These cement planter blocks eliminate the need to screw or glue, as one on each corner of the bed is sufficient to hold the raised walls into place. Thereafter, it can be filled with good-quality soil and planted. The advantages of planter blocks include that the beds can easily be disassembled to be moved or made bigger or smaller. The disadvantages of the system include that you will need to first properly prepare the ground beneath the bed to ensure it is level, so that the planter blocks and wooden frame lie flush against it, or you risk too much water draining out during irrigation.

If you would prefer a more permanent and aesthetic raised bed, consider a construction project. To build a bed that measures 1.2 by 2.4 meters, you will need three planks measuring 25 centimeters by 2.4 meters, and three smaller planks of 89 millimeters by 2.4 meters. Cut one of the three broader planks in half, to make the short edges of the bed (which are 1.2 meters long). If you do not have a saw, or are not comfortable cutting the wood yourself, you can have it cut to size at a local hardware store for a very reasonable fee.

Once the pieces are cut, arrange them into a rectangle in the area of your choosing, and then screw the corners together using four, evenly spaced 6 cm wood screws inserted into each corner, through the short edge of the wood. I recommend first pre-drilling a hole for the screws into the wood plank, to avoid it splitting.

Next, attach the thinner planks as a rim or lip around the bed, flat side down so that these planks are perpendicular to the ones below them. This helps the rigidity of the bed, and gives you an aesthetic finish, while also providing a ledge for you to place your tools or gardening equipment.

To make this project even cheaper, you can do without the thinner planks as an outside rim.

A PVC Pipe Vertical Garden

If you do not have the space for a raised bed, or you wish to have a vertical garden for ease or irrigation, you could consider upcycling a PVC pipe into a mini garden. Choose a thickness of pipe that will allow

148

for the most root space your plants require; pipes with a thicker diameter allow for more growing room. Using a drill and utility knife, cut out evenly spaced circles along the length of the pipe, through which your seedlings will emerge. Then, simply fill the entire pipe with soil, and plant your seeds or seedlings in the openings. Alternatively, you can have a horizontal garden, in which you could simply cut the entire length of PVC in half to create a trough, which can be planted and suspended from the roof or held in place by a stand.

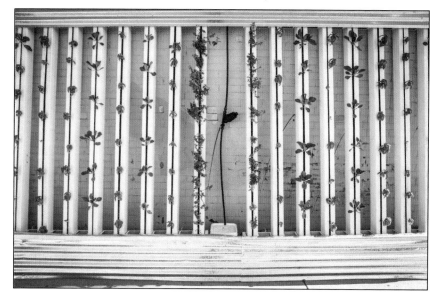

A Chicken Coop

To build an easy A-frame chicken coop which can comfortably house between three and five chickens, you will need three planks measuring 25 centimeters by 2.4 meters, and seven smaller planks of 8.9 centimeters by 1.8 meters, as well as two sheets of ply board measuring roughly 1.2 by 2.4 meters, a small bundle of chicken wire, a set of

149

Run from Easy   Chapter 8: Do it Yourself

hinges, some 6 cm wood screws, and a staple gun.

First, construct a box frame similar to the raised garden bed, using the three broader planks, but this time measuring 2.4 by 1.8 meters. To each of the shorter sides, and halfway along the length of the base, add two of the thinner planks measuring 1.8m which should be positioned to meet at the apex to form an A-Frame, and lay a final thinner plank horizontally across the three triangle tips. You should now have a three-dimensional triangle, resting on a flat rectangle. Using your staple gun, secure the chicken wire around the outside of the lower half of the frame. Thereafter, attach the plyboard sheets to each of the two sloping sides of the triangle, so that it resembles a roof. But cutting a square of this on one side and attaching hinges, you can easily enter the coop to retrieve eggs or add feed.

By adding wheels to the bottom base of the coop, you could make the structure mobile, which is a great way to use chicken waste as fertilizer, without chickens running loose across your property. The A—shape frame with slanting sides will protect your chickens from rain, and the plyboard top will provide them with shade.

Making a Composting Toilet

To create a very basic composting toilet, which lacks and frills and fuss, is easier than you'd imagine. You will need two 20-liter buckets of the same type (to ensure they have the same dimensions), four planks of wood that are cut to the same height as the buckets, a sheet of plywood, a toilet seat, and some screws and nails.

Start by cutting a hole in the plywood sheet that is as large as the rim of the bucket (as an easy measuring guide, place the bucket over the plywood and draw around the circumference with a pencil – anything inside the circle can be cut away. Once done, attach the toilet seat over the newly created hole. Then, use your planks to create legs for the toilet, by essentially creating a cube around the bucket. You should be able to remove the bucket easily from the frame. After attaching the plywood sheet with the toilet seat to the frame, the structure of your composting toilet is complete. All that is left to do is to add absorbents like sawdust or wood ash to the bucket, to neutralize smells. Ideally, you will need to swap out the buckets during use, so that one is used for wet waste, and the other for dry waste. This is the easiest way to create a rudimentary composting toilet that functions as a separator. Empty your buckets when they are three quarters full, or if the odor becomes unpleasant. For best results, locate the composting toilet in a cool, dry area that is not exposed to direct sunlight.

Cleaning Products

One of the most frequently overlooked aspects of self-sufficient living is cleaning, particularly if you plan to reuse your gray water. As such, it is worth knowing how to make your own cleaning products that are organic, non-toxic and environmentally friendly.

For example, a great recipe for a general all-purpose surface cleaner is to mix a cup of white vinegar with four teaspoons of baking soda and 20 drops of either tea tree or lavender essential oil. Be sure to use good-quality oils as these are added not just for their aroma, but for the disinfectant properties.

You can even make your own disinfectant wipes, which can thus be added to your compost heap if you have used a natural fabric such as cotton. To do so, soak your fabric squares in a mixture of one cup of water, eight drops of tea tree oil, eight drops of lavender oil, as well as ten drops of orange oil and a quarter of a cup of white vinegar. The added orange oil in this recipe makes for a great way to remove grease for hands.

For an organic dry carpet cleaner, mix two cups of borax with one cup of baking soda, and ten drops of essential oil. Not only is this a great way to clean your carpets, but borax also functions as an insect repellent and rat deterrent, so it is a double win!

How to Make Soap

Making your own soap is a nice skill to learn for a self-sufficient lifestyle. Not only can you avoid a trip to the store, but you will be able to wash without adding chemicals to your body or water. You could even consider making and selling soap to generate an income, in order to purchase other items that you cannot yet make yourself. To create a batch of soap, you will need 500 grams of coconut oil, 400 grams of ethically sourced palm oil, 600 grams of high-quality olive oil, 550 grams of distilled or purified water, approximately 250 grams of lye (sodium hydroxide, which can be purchased online or from a pharmacy) and seven drops of essential oil.

If you are making soap in your kitchen or using kitchen utensils, it is important to note that anything which makes contact with the lye should be stored separately and not reused while preparing or cooking food. In addition, because lye is caustic, it is a good idea to wear gloves and safety glasses while handling this ingredient. Once the soap-making process is complete, the lye is no longer toxic or caustic, as it undergoes chemical changes which render it totally harmless.

To begin, mix the lye. I would advise doing this outside if you cannot do it in a very well-ventilated indoor space. Weigh and set aside exactly 201 grams of lye. In a heat-safe vessel such as a glass jug, measure out your distilled water, then add the lye. Stir this mixture slowly and carefully, as mixing it will result in a chemical reaction that releases heat as a byproduct. This is why lye burns when it encounters your skin, as the moisture in skin triggers the reaction.

Once the lye has dissolved, set the mixture aside and allow it to cool to around body temperature. Depending on the climate and time of year, this usually takes between 30 - 90 minutes (Pike, 2020).

Once it has cooled, prepare a mold for your soap – this doesn't have to be fancy, you can use a lined baking tin (as long as you do not reuse it for baking in the future).

Next, melt all the oils together in a large container – you can do this on the stovetop or in a microwave – so that they are a liquid with even consistency. Next, add your essential oil mixture, and then allow the oil mixture to cool to just about room temperature. Once your oils and your lye mixture are at the same temperature, you can blend these by adding the lye to the oils, and not the other way around (to avoid any reactions). Once these two ingredients mix, they will turn cloudy. Then, whisk the soap batch or use a stick blender to ensure they are properly blended. Ideally, keep mixing for between 3 - 5 minutes, allowing the mixture to thicken. Finally, pour the mixture into your mould, and cover the mixture with a sheet of cling wrap leaving it to settle for 24 hours. Then, cut the soap into blocks and allow them to cure for between four and six weeks. This curing process is essential, as it allows any water in the mixture to evaporate, making it both physically harder but also milder on your skin.

Shopping & Clothing

With regards to clothes, it is not very easy to become self-sufficient from purchasing all together. However, in the last 10 years there has been a resurgence in knitting among the younger and older demographic. Another skill worth re-learning is sewing; how to use a sewing machine to make your own clothes. Even with both of these skills in your arsenal, you are still left with the problem of sourcing the materials. Unless you are keeping sheep and desire to undertake the process of sending your shorn wool to be processed into yarn, you will likely need to purchase materials. Whether you pay someone to process your wool or purchase the materials from the shop directly, costs are involved. Personally, I think it would be too cumbersome in time and effort to expect to be fully independent in in this area. You are better off selling the wool produced from your sheep and purchasing some materials to grow your sewing and knitting skills; but also, not to expect to have all your clothing needs met by this.

My suggestion to those who live in the city or don't own sheep is this: buy second-hand clothing from charity shops to bring down your cost of clothes shopping significantly. It is very possible to find good quality, branded clothes in charity shops for as low as £1. Once you get the smell for a bargain, you never go back. Remember, if you budget your clothing expenses in such a way, that your savings/earnings from other self-sufficient activities compensate for your shopping expenditure, you can consider yourself to be as self-sufficient as possible. This is true of all of your purchasing habits, not just for clothing.

Furthermore, if you did want to experiment in learning skills such as sewing and knitting, I would recommend trying to reuse old clothes that are no longer useable or don't fit and turning them into something else e.g., a scarf or gloves etc. If you become proficient in these skills, they could be another means of income for you, selling products with a sustainable, eco-friendly edge to them, i.e., selling hand-sewn clothes made from upcycled materials and secondhand clothing…

Old clothing items can also find new use in the garden. For example, using clothes as protective meshing to cover plants or growing areas, they can also be used to filter water, some clothing made of 100% natural biodegradable materials, can even be added to the compost heap.

If at First You Don't Succeed, Try Again

The most important part of learning new skills is to enjoy them. Without a passion for a self-sufficient lifestyle, DIY will simply become a chore. Try to enjoy the process of learning and growing in these various skills, as they are your key to a self-sufficient lifestyle. After all, no one can master skills or new challenges overnight, and you may require several attempts to master a new skill. Do not let this deter you. With time and patience, learning new skills can be more rewarding, and anything you initially struggle with will get easier with time. As with everything in life, self-sufficient lifestyles are a learning curve, but if you keep at it, you will become efficient at jobs you once thought you could not do.

Conclusion

Sadly, we have reached the end of our journey together. Hopefully, you're no longer intimidated by the idea of self-sufficient living. Instead, I hope you are excited by the sheer number of ways that an off-the-grid lifestyle could benefit you, your family, and the environment.

At the start of this book, I wanted to ensure that as you continued to read, you would learn enough about independent living that the subject would no longer be foreign nor the lifestyle unattainable, and I truly hope that I have succeeded.

The main ideas that I have tried to share with you are as follows:

1. Anyone can start living more independently – regardless of the size and composition of your home and family.

2. There are different levels of self-sufficiency. Independent living is nuanced. There is no need to embrace extreme off-the-grid living if you simply want to implement a few ideas to reduce your bills and be more environmentally friendly. How far along the journey you travel towards independence up to you.

3. By applying the techniques, I have introduced; you will be able to efficiently start a few basic ways of becoming independent, without falling victim to any of the common mistakes.

4. A self-sufficient lifestyle is good for your wallet, your physical health, and the environment. It is also good for your mental health as you feel less a slave to the 9 – 5. It provides you with more choice because you aren't relying on it for provision as heavily as you once were.

5. This is a process that you should enjoy from start to finish, even though it may be hard to make some of the required lifestyle changes. The benefits far outweigh the sacrifice.

6. There's no quick fix for exiting a life defined by consumer culture – it requires time and patience. It takes work, but with practice, it will get easier

If you enjoyed this book, please let me know by leaving a review on Amazon – your feedback helps me not only in my own journey towards self-sufficiency, but in writing, too. Thank you for choosing me to help you master your transition to an independent lifestyle, and I wish you everything of the best in your home journey.

Self Sufficiency Tracker

Now that we have come to the end of this journey together, I recommend you track your progress. Self-sufficiency is not an overnight thing but requires time and patience and the best way to reach independence is one step at a time. However, if you don't take action with the smaller, easier tasks, you will not be able to achieve your bigger dreams of independence. You might start simple by reducing your bills or by growing a few vegetables in your garden or by starting a compost heap. Either way it is step toward your goals. It can be hard to feel the benefits of the progress you have made in a real monetary way. That is why I created a simple planner that you can use to track your progress over time. And guess what, it is completely free! Hopefully it inspires you

with ideas you can use to either generate self-sufficient income to increase the percentage of your lifestyle that is self-sufficient or ideas to reduce your dependence on convenience. The link to get sent a copy of the planner is: vaughanpublishing.activehosted.com/f/1 Also, keep an eye out for my other books to come to do with green lifestyle and how we can live a less toxic and healthier life both for us, the community and the planet.

References

Alterman, T. (n.d.). *How do I pasteurize raw milk at home?* Mother Earth News. Retrieved June 10, 2021, from https://www.motherearthnews.com/real-food/pasteurize-raw-milk-at-home

Aquavision. (2021). *Aquaponics and the law.* Www.aquavisiononline.com. https://www.aquavisiononline.com/index.php?page=aquaponics-and-the-law

Amber, M. (2019). water well tiled roof. In Pixabay. https://pixabay.com/photos/water-well-tiled-roof-bucket-4247735/

Australian Government. (2010). *Wastewater reuse.* Yourhome.gov.au. https://www.yourhome.gov.au/water/wastewater-reuse

Barth, B. (2018, October 19). *How does aeroponics work?* Modern Farmer. https://modernfarmer.com/2018/07/how-does-aeroponics-work/

BBC. (2020, May 6). *How to grow vegetables – beginner veg to grow.* BBC Gardeners' World Magazine; https://www.gardenersworld.com/plants/vegetable-crops-for-beginners/

BEE-health. (2019, August 20). *Basic beekeeping techniques.* Bee-Health.extension.org; US Department of Agriculture. https://bee-health.extension.org/basic-beekeeping-techniques/

Bessa, R. (2019, April 12). *Raising pigs: a pig farming guide for beginners.* Farm4Trade. https://www.farm4trade.com/raising-pigs-a-pig-farming-guide-for-beginners/

Bjarnadottir, A. (2019, March 8). *Beetroot 101: Nutrition facts and health benefits.* Healthline; Healthline Media. https://www.healthline.com/nutrition/foods/beetroot

BrightAgrotech. (2015). Hydroponic garden. In *Pixabay.* https://pixabay.com/photos/hydroponics-green-wall-zipgrow-917285/

Britannica. (n.d.). *Food preservation.* Encyclopedia Britannica. Retrieved June 11, 2021, from https://www.britannica.com/topic/food-preservation/Light-induced-reactions#ref50552

Campion, A. (2020, June 18). *Basic tools that everyone should have.* Www.confused.com. https://www.confused.com/home-insurance/guides/tool-kit-list-for-every-home

Clarke, S. (2018, December 7). *Vehicle Conversion to Natural Gas or Biogas.* Www.omafra.gov.on.ca. http://www.omafra.gov.on.ca/english/engineer/facts/12-043.htm

Countryside. (2019, November). *Sheep farming for beginners and beyond.* I Am Countryside. https://www.iamcountryside.com/freeguide/sheep-farming-for-beginners-and-beyond/

CSGN. (n.d.). Garden bed types: advantages and disadvantages. In *Collective Schools Garden Network.* Western Growers Foundation. Retrieved June 6, 2021, from http://www.csgn.org/sites/csgn.org/files/Garden%20Bed%20Types.pdf

Department of Agriculture. (2017, January 26). *Small-scale egg production.* Farmer's Weekly; Caxton Magazines. https://www.farmersweekly.co.za/farm-basics/how-to-livestock/small-scale-egg-production/

Driessen, S. (2020). *Drying food at home.* Extension Food Safety; University of Minnesota. https://extension.umn.edu/preserving-and-preparing/drying-food

Energy Saving Trust. (2021). *Use hydroelectricity to power your home.* Energy Saving Trust. https://energysavingtrust.org.uk/advice/hydroelectricity/#:~:text=the%20average%20home.-

Ershova, K. (2021).

Greenhouse planting spring. In Pixabay. https://pixabay.com/photos/greenhouse-planting-spring-beds-6226263/

Facility Management. (2021). *How much wind is needed for a small wind turbine?* Www.calgary.ca. https://www.calgary.ca/cs/cpb/operations-workplace-centre/bearspaw-owc/wind-need-for-small-turbine.html

Farmer's Weekly. (2017, March 14). *How to develop and manage a small beef cattle herd.* Farmer's Weekly; Caxton Magazines. https://www.farmersweekly.co.za/farm-basics/how-to-livestock/develop-manage-small-beef-cattle-herd/

Fong, J., & Hewitt, P. (2019). *Worm composting basics*. Cornell.edu. http://compost.css.cornell.edu/worms/basics.html

Girardi, A. (2013, July 24). *The pros and cons of installing a water well in your home*. Totally Home Improvement. https://www.totallyhomeimprovement.com/exterior/installing-home-water-well

Godfrieda, M. (2009). solar panels placement. In Pixabay. https://pixabay.com/photos/solar-panels-placement-green-energy-944000/

Goellner, A. (2019). crop celery field. In Pixabay. https://pixabay.com/photos/crop-celery-field-agriculture-4230045/

High Mowing Organic Seeds. (2015, May 7). *Our top 10 crops for beginner gardeners*. Www.highmowingseeds.com. https://www.highmowingseeds.com/blog/successful-vegetables-our-top-10-crops-for-beginner-gardeners/

Hughes, P. (2019, March 1). *50 simple ways you can reduce emissions and cut bills*. Inews.co.uk. https://inews.co.uk/news/environment/how-to-reduce-your-energy-consumption-50-simple-ways-to-cut-bills-and-reduce-emissions-264785

James, C. (2019, September 5). *Becoming self-sufficient: how and why to do it*. ABC Money. https://www.abcmoney.co.uk/2019/09/05/becoming-self-sufficient-how-and-why-to-do-it/#:~:text=Everyone%20has%20a%20different%20reason

Just Energy. (2021). *Hydropower 101*. Justenergy.com. https://justenergy.com/blog/hydropower-101/

Keeper of the Home. (2017, January 16). *The ultimate guide to homemade all-natural cleaning*. Keeper of the Home.

https://keeperofthehome.org/homemade-all-natural-cleaning-recipes/

Kerckx, B. (2014). green compost waste. In Pixabay. https://pixabay.com/photos/green-waste-compost-compost-bin-513609/

Kubala, J. (2017, November 4). *9 Impressive health benefits of cabbage*. Healthline. https://www.healthline.com/nutrition/benefits-of-cabbage#TOC_TITLE_HDR_2

Kubala, J. (2018, December 18). *9 Impressive health benefits of onions*. Healthline. https://www.healthline.com/nutrition/onion-benefits#TOC_TITLE_HDR_5

Kuptsova, A. (2016). pickels billet cucumbers. In Pixabay. https://pixabay.com/photos/pickles-billet-cucumbers-1799731/

Ladonski, A. (2019, August 26). *Organic pest control methods*. Extension.sdstate.edu. https://extension.sdstate.edu/organic-pest-control-methods

Marcus, S. (2020, December 21). *The cheapest, most efficient way to heat your home this winter | OVO Energy*. Www.cms-App-Prod.ovotech.org.uk. https://www.ovoenergy.com/guides/energy-guides/is-it-more-energy-efficient-to-leave-the-heating-on.html

MarsRaw. (2019). hydroponics greenhouse. In Pixabay. https://pixabay.com/photos/hydroponics-greenhouse-lettuce-4255403/

Martin, A. (2019, September 23). *The top 14 healthiest greens for your salad*. EverydayHealth.com. https://www.everydayhealth.com/diet-nutrition-pictures/best-salad-greens-for-your-health.aspx

Maxwell-Gaines, C. (2004, April 4). *Rainwater harvesting 101*. Innovative Water Solutions LLC; https://www.watercache.com/education/rainwater-harvesting-101

McDougal, T. (2020, June 5). *US/UK antibiotic use in poultry examined*. PoultryWorld. https://www.poultryworld.net/Health/Articles/2020/6/USUK-antibiotic-use-in-poultry-examined-593547E/

Mclaughlin, R. (2020, March 2). *How to preserve meat with salt*. Delishably. https://delishably.com/meat-dishes/How-to-Preserve-Meat-with-Salt

Millar, A. (2018, March 19). *Watercourses, boundaries and riparian rights – what landowners need to know*. Tallents Solicitors. https://www.tallents.co.uk/riparian-rights-what-you-need-to-know/

Muller, C. (2021). *Basic principles of dairy farming*. South Africa. https://southafrica.co.za/basic-principles-dairy-farming.html

National Hydropower Association. (2010). *Affordable - National Hydropower Association*. National Hydropower Association. https://www.hydro.org/waterpower/why-hydro/affordable/

National Parks Association. (n.d.). *How to collect water*. National Parks Association of NSW. Retrieved June 15, 2021, from http://www.bushwalking101.org/how-to-collect-water/

Neverman, L. (2018, April 4). *How to can food at home*. Common Sense Home. https://commonsensehome.com/how-to-can-food-at-home/

North, D. (2016, May 30). *What is aquaponics and how does it work?* The Permaculture Research Institute. https://www.permaculturenews.org/2016/05/30/what-is-aquaponics-and-how-does-it-work/

Oetken, N. (2016, June 8). *How to easily dig a well on your property*. Survival Sullivan. https://www.survivalsullivan.com/how-to-dig-a-well/

Oliver, J. (2017). water wheel pump rio. In Pixabay. https://pixabay.com/photos/wheel-water-pump-rio-nature-2670662/

OVO Energy. (2019). *OVO vehicle-to-grid charger*. Ovoenergy.com. https://www.ovoenergy.com/electric-cars/vehicle-to-grid-charger

Pike, C. (2020, August 12). *How to make soap at home*. Food52. https://food52.com/blog/12919-how-to-make-soap-at-home-even-if-you-failed-chemistry

Ploetz, K. (2013, October 28). *How do I know if backyard slaughter is legal?* Modern Farmer. https://modernfarmer.com/2013/10/dear-modern-farmer-know-backyard-slaughter-legal/

Renewable Energy Hub. (2010). *Legal plan and permission wind turbines*. The Renewable Energy Hub. https://www.renewableenergyhub.co.uk/main/wind-turbines/legal-planning-permission-for-wind-turbines/

RunnerDuck. (n.d.). *Waterwheel project page*. Www.runnerduck.com. Retrieved June 18, 2021, from http://www.runnerduck.com/wheel10.htm

Sakawsky, A. (2017, June 28). *Pickling 101*. The House & Homestead. https://thehouseandhomestead.com/ultimate-guide-to-pickling/

164

Sass, C. (2018, August 2). *7 Health benefits of tomatoes*. Health.com. https://www.health.com/nutrition/health-benefits-tomatoes

Scheckle, P. (2018). *Make a Biogas Generator to Produce Your Own Natural Gas | MOTHER EARTH NEWS*. Mother Earth News. https://www.motherearthnews.com/renewable-energy/other-renewables/biogas-generator-zm0z14aszrob

Scoob, M. (2019). switzerland cheese dairy. In Pixabay. https://pixabay.com/photos/switzerland-cheese-dairy-craft-4598579/

Sengupta, S. (2018, July 19). *8 Incredible benefits of peas*. NDTV Food. https://food.ndtv.com/food-drinks/8-incredible-benefits-of-peas-you-may-not-have-known-1798358

Smith, N. (n.d.). How to make corn flour. LIVESTRONG.COM. Retrieved June 19, 2021, from https://www.livestrong.com/article/13729094-weight-loss-program-testimonials-noom/

Stafford, G. (2018, July 22). *How to make homemade butter*. Gemma's Bigger Bolder Baking. https://www.biggerbolderbaking.com/homemade-butter/

Successful Farming. (2020, February 7). *When and how to start a poultry farm*. Agriculture. https://www.agriculture.com/livestock/poultry/when-and-how-to-start-a-poultry-farm

Taste. (2020). *How to make home-made yoghurt*. Woolworths TASTE. https://taste.co.za/how-to/how-to-make-home-made-yoghurt/?gclid=Cj0KCQjw8IaGBhCHARIsAGIRRYpqU-iZqBr_SRRBeZ_E7hdp7t2QtaNpwpj4q6-2S7RkW1olI3HMrBsaAnKhEALw_wcB

The Bug Out Bag Guide. (2014, September 10). *What Is bushcraft?* The Bug out Bag Guide. https://www.thebugoutbagguide.com/what-is-bushcraft-survival/

The Milk Maid. (2015, January 14). *Basic steps of how to make cheese*. Instructables; https://www.instructables.com/Basic-Steps-of-How-to-Make-Cheese/

The Renewable Energy Hub. (2018). *How much does a wind turbine cost?* The Renewable Energy Hub. https://www.renewableenergyhub.co.uk/main/wind-turbines/how-much-does-a-wind-turbine-cost/

US EPA. (2018, October 16). *Composting at home.* US EPA. https://www.epa.gov/recycle/composting-home

US EPA. (2019, March 18). *How does anaerobic digestion work?* US EPA. https://www.epa.gov/agstar/how-does-anaerobic-digestion-work

Van Buuren, E. (2020, August 31). *10 Benefits of hydroponics.* Eden Green Technology. https://www.edengreen.com/blog-collection/benefits-of-hydroponics

Vanderlinden, C. (2019, March 12). *How composting how-to guide.* The Spruce. https://www.thespruce.com/how-to-hot-compost-2539474

Vekony, A. T. (2021, May 31). *How do solar panels work?* Www.greenmatch.co.uk. https://www.greenmatch.co.uk/solar-energy/solar-panels#how-they-work

White, A. (2020, November 4). *Updated a-frame chicken coop.* Www.ana-White.com. https://www.ana-white.com/woodworking-projects/updated-frame-chicken-coop-plans

WooWoo Waterless Toilets. (2021). *What is a compost toilet?* WooWoo Waterless and Composting Toilets. https://www.waterlesstoilets.co.uk/compost-toilets/

Printed in Great Britain
by Amazon